Reproduction ...a
in the Pig

Reproduction in the Pig

P. E. Hughes, BSc, PhD, MIBiol
Department of Animal Physiology and Nutrition, University of Leeds, UK

M. A. Varley, BSc, PhD, MIBiol
Department of Applied Nutrition, The Rowett Research Institute, Aberdeen, UK

The Butterworth Group

United Kingdom	**Butterworth & Co (Publishers) Ltd** London: 88 Kingsway, WC2B 6AB
Australia	**Butterworths Pty Ltd** Sydney: 586 Pacific Highway, Chatswood, NSW 2067 Also at Melbourne, Brisbane, Adelaide and Perth
Canada	**Butterworth & Co (Canada) Ltd** Toronto: 2265 Midland Avenue, Scarborough, Ontario, M1P 4S1
New Zealand	**Butterworths of New Zealand Ltd** Wellington: T & W Young Building, 77–85 Customhouse Quay, 1, CPO Box 472
South Africa	**Butterworth & Co (South Africa) (Pty) Ltd** Durban: 152–154 Gale Street
USA	**Butterworth (Publishers) Inc** Boston: 10 Tower Office Park, Woburn, Mass. 01801

First published 1980

ISBN 0 408 70946 4 (Cased) 0 408 70921 9 (Limp)

British Library Cataloguing in Publication Data

Hughes, P.E.
Reproduction in the pig.
1. Swine – Reproduction
I. Title II. *Varley*, M.A.
636.4′08′926 SF396.9 80-40241

ISBN 0-408-70946-4
ISBN 0-408-70921-9 Pbk

Typeset by Scribe Design, Gillingham, Kent
Printed and Bound in England by Redwood Burn Ltd
Trowbridge & Esher

Preface

In recent years much effort has been directed towards the gaining of a clearer understanding of reproductive processes in the pig. Many areas of study have been included in this work. In particular the fields of endocrinology and physiology have made significant contributions to our knowledge. It is the purpose of this review to integrate the work into one text covering all the related disciplines, in order to establish both the current state of knowledge and the gaps therein. The ultimate aim of this endeavour is the elucidation of the way ahead for the commercial pig industry, and the stimulation of increased efficiency in production. Hence this book is aimed at persons engaged in both the advisory and teaching professions associated with the pig industry. In particular it is hoped it will provide an adequate 'base camp' for individuals entering or already involved in pig reproduction research. The authors also hope that the book will be read by progressive farmers, and students about to embark on a career in the pig industry. With this in mind each chapter includes a general concluding section to give the practical implications of its text.

It is recognized that terminology in the pig industry is slightly different in different nations and a glossary of terms has been included to provide explanation for frequently used common terms. In addition, a table of conversion factors for metric and imperial measures is given.

The book begins with an analysis of the cerebral and endocrine control systems involved in reproduction and carries on to review the prepubertal stage in the pig. This leads into the reproduction of the mature sow with chapters on the oestrous cycle, ovulation, fertilization and conception, pregnancy, litter size, lactation and the weaning to remating interval. The reproductive mechanisms in the boar are then dealt with under a separate section. This includes chapters on puberty in the male, fertility, mating behaviour and artificial insemination. Chapter 15 then gives a concise summation of the authors' practical recommendations in the light of the preceding chapters.

Hence it is hoped the format used provides a logical sequence of presentation of factual information and interpretation of this information.

The authors are indebted to the many people who kindly gave permission for the use of illustrations and graphic data, and specific acknolwedgement is included where relevant in the text. We would also like to thank both Mrs Caroline Varley and Mrs Sylvia Bennett for painstaking work in typing the original manuscript. In addition we would like to gratefully acknowledge the guidance over the years of our mentor Dr Des Cole.

M.A. VARLEY
P.E. HUGHES

Contents

Chapter 1

Introduction

Fertility among apparently normal sows living under what would seem to be normal conditions is less than it could be.
J. A. Laing (in 1957)

Fertility in the pig is certainly a character with a high degree of variability and this is also true of all the components of the reproductive biology of the species. It would appear that John Laing had the feeling that in his day there was much progress to be made in the exploitation of the pig's capacity for prolificacy and fecundity. This vast differential between theoretical potential and realized output still exists today despite a wealth of knowledge accrued since 1960 in the field of the reproductive biology of the pig. It is hoped therefore in the ensuing chapters that answers can be found to explain this observation and a clearer understanding of the mechanisms involved is the outcome.

1.1 Reproductive output

Given two litters a year and a litter size somewhat over ten piglets born alive it should be possible to produce about 20 saleable pigs from each sow every year. The number of piglets produced per sow per year is a major factor influencing the profitability of pig production. Sow feed can be considered as a fixed cost and the more piglets this fixed cost is spread over, the cheaper is the cost of producing each piglet and the greater is the margin left for the farmer.

Any reduction of annual sow output therefore leads to an increase in costs and nationally represents inefficient use of a scarce food resource resulting in higher pig meat prices to the consumer.

Table 1.1 presents the actual level of sow productivity being achieved by various pig producing nations. It can be seen that there is a large discrepancy for all of these nations between the target of 20 pigs produced per sow and the productivity actually achieved. Clearly there is a lot of room for improvement and many farmers are not efficiently utilizing the resources at their disposal. Much of this wastage must be the result of a lack of understanding of the reproductive processes of the pig.

Table 1.1. Annual sow productivity*

Country	*Sow numbers* (1000s)	*Total slaughter pigs produced* (1000s)	*Sow productivity: (slaughtered pigs produced/sow/year)*
United Kingdom	886	12 881	14.5
W. Germany	2329	31 785	13.6
Netherlands	945	10 654	11.3
Italy	896	7 687	8.6
Irish Republic	116	1 726	14.8
France	1428	15 793	11.1
Denmark	1022	10 329	10.1
USA	8000	74 250	9.4

*Figures relate to 1976 (M.L.C. International Market Survey, 1977)

Table 1.2. The effect of sow productivity on the profitability of weaner production*

No. of weaners reared/sow/year	22	20	18	16	14
	%	%	%	%	%
Feed costs/sow	100	98.4	96.9	95.4	93.8
Total costs/sow	100	98.9	97.9	96.8	95.8
Sales	100	90.8	81.8	72.7	63.6
Gross margin/sow	100	82.2	65.1	48.0	30.8

*From M.L.C. Feed Recording Service Report No. 2 (1975)

The Meat and Livestock Commission (M.L.C.) have tried to quantify in financial terms what the effect of a change in sow output is on profitability and the results are given in *Table 1.2.*

There is a very marked change in profitability as sow output drops from 22 to 14 and it is worth noting that these figures relate to 1974, a fairly healthy economic climate for pig producers. If the same analysis had been carried out in the following years much of the data for margin/sow would have been in a loss situation except for the ones with the highest annual sow productivity.

1.2 Components of reproduction

The productivity of a breeding herd is dependent on the annual output from the sow. This annual sow productivity has in fact two main components: Litter size being the first and the farrowing index which is the number of litters farrowed per sow per year being the second. These in turn have various components contributing to their expression and these are shown in *Figure 1.1.* It may be that in trying

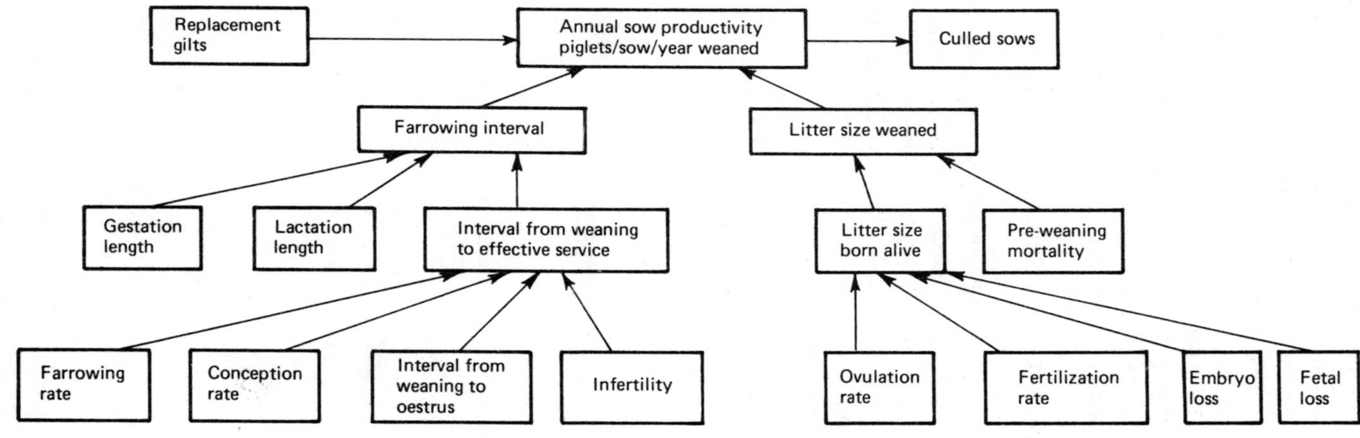

Figure 1.1 The components of sow productivity

to close the gap between theoretical and observed sow productivity attention should be focused on some of these components more than others. It is hoped therefore that by the end of this book the reader will have seen what are the biological relationships controlling these components, what are the limitations to further improvement and which components are the most ripe for development in the future.

For example, gestation length is a prime component of farrowing interval (the average time for all sows in a herd between successive farrowings) but this factor is a constant component out of the scope of influence by management whereas lactation length which is also a major influence on farrowing interval is directly under management control and could potentially offer tremendous progress in improving annual sow productivity.

We shall therefore review these important components of the performance of the breeding herd taking the boar and the sow through the prepubertal period and on to maturity and 'normal' reproductive life and shall include both research conclusions and practical interpretation to give an integrated picture of this complex and fast growing area of knowledge.

Chapter 2

The basis of reproduction

It is of little value considering how we may exert influence over the reproductive processes of the female if we do not first consider the normal events that occur, together with their control mechanisms. In addition, the advantages of increased reproductive potential in the female are lost if fertility in the male is low. The purpose of this chapter is to provide a basic knowledge of the control of reproduction in both the male and female pig, upon which we may build in later chapters.

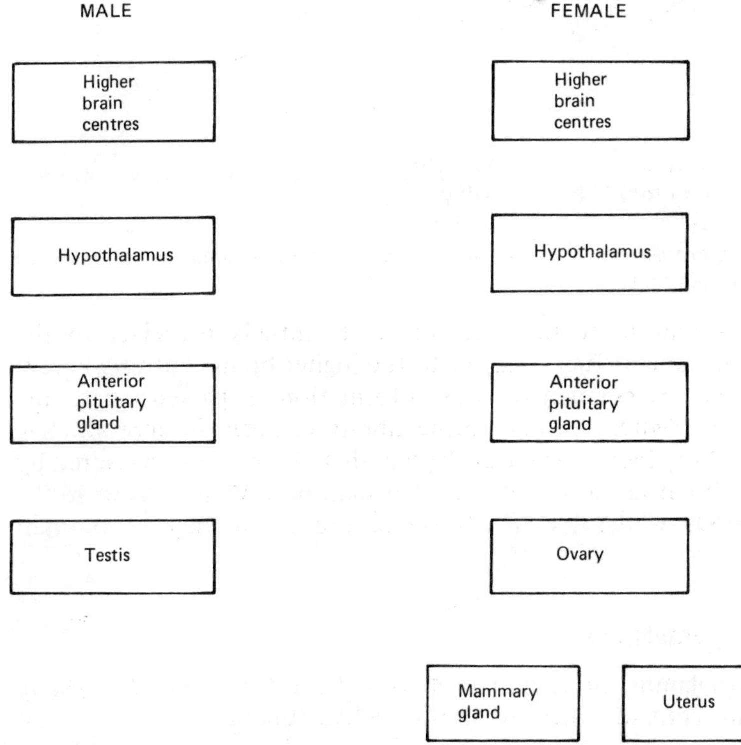

Figure 2.1 The primary tissues involved in reproduction

To begin with, we must identify those tissues which are of primary importance to the reproductive process. These are considered to be the higher brain centres, hypothalamus, anterior pituitary gland, gonads and accessory structures. These tissues are diagramatically represented in *Figure 2.1.*

2.1 The higher brain centres

The higher brain centres include the pineal gland, olfactory lobes and all other central nervous system areas (excluding the hypothalamus) that are involved in the coordination and integration of information on the environment. These centres are shown in *Figure 2.2*. The

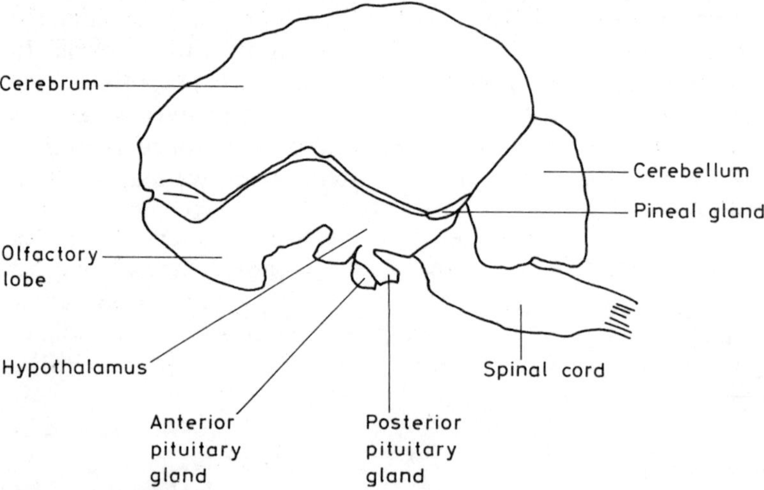

Figure 2.2 A cross-section of the pig brain showing those areas oı relevance to reproductive function

information relating to the environment is initially perceived by the sense organs, and is then relayed to the higher brain centres where it is monitored. Subsequently, this information is passed on to the hypothalamus where it may bring about changes in reproductive function. Thus, factors such as daylength and stress are measured by the higher brain centres, and this information is then relayed to the hypothalamus where its effects on reproduction may be brought about.

2.2 The hypothalamus

The hypothalamus, in conjunction with the anterior pituitary gland, is the main control centre for reproductive function. It is here that information on both the internal and external environment is translated into the appropriate output for the reproductive system. As has already been mentioned, information on the external environment is primarily received from the higher brain via internal brain pathways. Information on the internal environment is obtained by monitoring blood levels of both reproductive and metabolic hormones at the hypothalamus.

Once the information has been received by the hypothalamus it must be translated into the appropriate reproductive output. This is via variations in the level of secretion of hypothalamic hormones. These hypothalamic hormones occur in two forms, known as releasing hormones (RHs) and inhibiting hormones (IHs). They are produced in the hypothalamus and, following release, are carried to the pituitary gland by the hypophysial portal blood vessels. Once at the pituitary gland the releasing hormones, as their name might suggest, stimulate the release of pituitary hormones into the circulation.

2.3 The anterior pituitary gland

This gland is situated immediately below the hypothalamus, and is attached to the anterior side of the posterior pituitary gland. Its chief function is the synthesis, storage and secretion of reproductive and metabolic hormones. Although the pituitary gland produces at least eight different hormones, only three are of importance to the present discussion. These are follicle stimulating hormone (FSH), luteinizing hormone (LH) and prolactin (PL).

All three of these hormones may be released into the circulation in response to the arrival of hypothalamic releasing hormones. In the case of FSH and LH it is thought that the releasing hormone is common to both hormones. However, prolactin is also known to have an inhibiting hormone and this is considered to be the predominant hypothalamic controller of prolactin. FSH and LH releasing hormones will therefore be called gonadotrophin releasing hormone (GnRH) and prolactin inhibiting hormone will be called PIH.

2.4 The female: The gonads and accessory structures

THE OVARY

In the female the primary functions of the two pituitary hormones FSH and LH are the stimulation of follicular growth and the subsequent release of the mature ova at ovulation. It is now recognized that both hormones are required in each of these processes. However, the predominant role of FSH is the stimulation of follicular growth and development, whereas LH is predominantly involved with ovulation. In fact, the normal ovarian changes which occur in response to gonadotrophic stimulation (i.e. by the two gonadotrophic hormones FSH and LH) may be considered in three stages. First, the ovarian follicles grow and develop in response to high circulating levels of FSH in the presence of LH. Next, once these follicles have fully developed, they shed their mature ova. This occurs in response to high blood levels of LH, although recent evidence suggests that a small elevation of FSH is also necessary at this stage. Finally, after the ova have been shed (i.e. once ovulation has occurred), the ruptured follicles from which the ova have been released form corpora lutea. This final step is controlled by the presence of low circulating levels of the gonadotrophins.

In addition to these gross changes, the ovary also synthesizes and secretes its own hormones, namely oestrogens and progesterone. These are steroid hormones produced by the developing follicles and corpora lutea respectively. Thus, the output of oestrogen reaches a maximum just prior to ovulation, when follicular growth is at its greatest. Progesterone, on the other hand, attains a maximum rate of release once corpus luteum formation is complete.

THE UTERUS

At ovulation mature ova are shed into the fallopian tubes and pass down into the uterus. If mating occurs at this time and the ova are fertilized, then fertilized ova will develop and implant within the uterus and a pregnancy will ensue. However, the uterine conditions required for mating and for pregnancy are quite different. At the time of mating the uterus must supply optimum conditions for ovum and sperm transport, whereas during pregnancy uterine conditions must be those necessary for the implantation and development of the embryo.

These uterine changes are brought about by the ovarian steroid hormones. In the period immediately preceding ovulation increasing amounts of oestrogens are being secreted by the developing follicles. These act on the uterus to provide optimum conditions at the time of mating (it should be noted at this point that oestrogen is also primarily responsible for the behavioural oestrus, i.e. the standing heat reflex). Following ovulation, progesterone becomes the dominant ovarian hormone. This then influences uterine development, providing optimum conditions for pregnancy.

THE MAMMARY GLAND

Once a pregnancy is complete and the young are born, the mother produces milk in order to feed her offspring. It is at this stage that prolactin (together with growth hormone) becomes the dominant

Issues Summary 17/04/2023 10:51

Writtle College Library

Id :: 143497****
Name:: Jessica Goldsmith

Item : 000188342 , Reproduction in the pig
08/05/2023

Total Number of Issued Items **1**

Thank you for using Self-Service

secretion from the anterior pituitary gland. In fact, prolactin appears to be released from the pituitary gland in response to the suckling stimulus. Once secreted into the circulation prolactin acts at the mammary gland to stimulate milk production. (It should be added that the synthesis and secretion of milk does involve many other hormones. However, since prolactin is the primary lactation hormone concerned with reproduction, it is considered unnecessary to discuss the mechanisms of lactation in detail at this stage.)

2.5 The male

THE TESTIS

The testis is composed of two major tissue types, these being the Leydig cells and the seminiferous tubules. The primary function of these two tissues are, respectively, the synthesis and secretion of the male gonadal hormone testosterone, and the development of mature spermatozoa from primitive germ cells (a process known as spermatogenesis). Testosterone production is mainly in response to LH stimulation, although FSH may also play a minor role. The testosterone secreted in response to this gonadotrophic stimulation itself has a stimulaory effect on spermatogenesis. Indeed, the presence of testosterone is essential for several stages of the sperm maturation sequence, whereas other steps in this sequence require the presence of FSH. Thus, the gonadotrophins stimulate the production of testosterone, and this hormone (plus FSH) stimulates the development of mature spermatozoa.

2.6 Control systems

At this stage we have a system in which the hypothalamus secretes releasing and inhibiting hormones, the anterior pituitary gland releases gonadotrophic hormones (i.e. FSH and LH) and prolactin, and the gonads secrete the steroid hormones oestrogen, progesterone and testosterone. This system is shown in *Figure 2.3*. However, it is clear from this figure that the picture is not complete. The problem is that all the steps, with the exception of PIH, are stimulatory and, hence, hormone release would continually increase and thereby be uncontrolled. The fact that circulating hormone levels are not normally excessively high is due to two factors: firstly the restraining influence of the external environment on the production and release of releasing hormones and, secondly, the restrictions imposed by the internal environment.

The influence of the external environment is not a primary control mechanism in reproductive function. Its role may be considered as

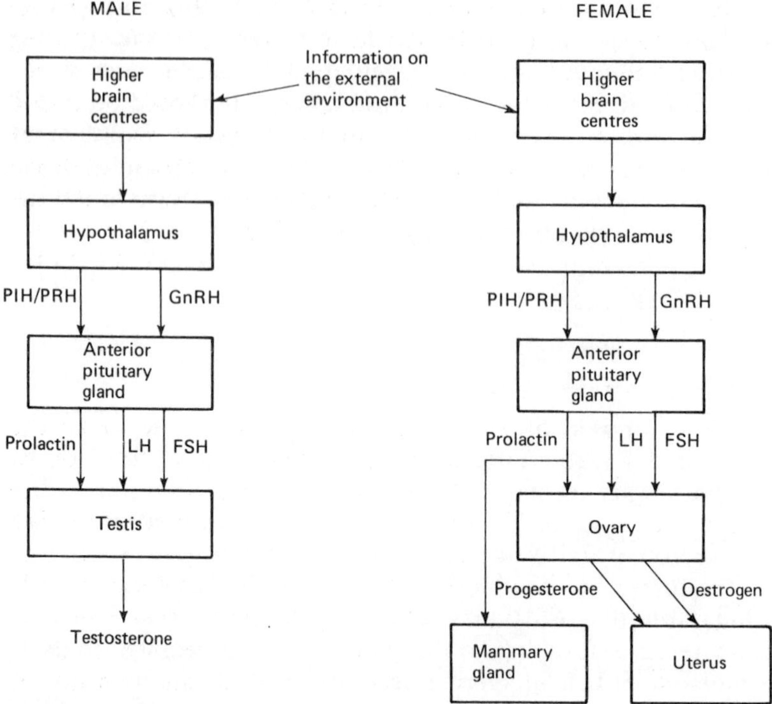

Figure 2.3 A diagrammatic representation of the hormones secreted by those tissues primarily involved in reproduction

that of a regulator of reproductive activity in accordance with the suitability, or otherwise, of environmental conditions for the production of offspring. Thus for example, fertility in both the sow and the boar declines during the summer months. This results in fewer litters being born during the cold winter months when, under natural conditions, piglet survival would be reduced by both food shortage and adverse weather conditions.

In contrast, the restrictions imposed by the internal environment are considered to constitute the primary controlling mechanism in reproduction. This control is achieved by the monitoring of circulating reproductive and metabolic hormone levels at the hypothalamus and pituitary gland via mechanisms known as feedback loops.

THE FEEDBACK LOOPS

The term feedback loop is used to describe a mechanism whereby the brain monitors the circulating levels of certain hormones and regulates their rate of secretion. For example in the female the hypothalamus secretes GnRH which promotes the release of FSH and LH from the anterior pituitary gland. These are then carried, via the circulation, to the ovary where they stimulate the production and secretion of oestrogen. At this stage, if no feedback loop was operative, the rate of production and release of oestrogen would continually increase. However, in practice oestrogen forms a negative feedback loop with the hypothalamus. This means that circulating levels of oestrogen are monitored at the hypothalamus such that high oestrogen levels reduce the rate of secretion of GnRH whereas low oestrogen levels allow GnRH release to continue. In this way blood levels of FSH, LH and oestrogen remain stable at relatively low levels (*see Figure 2.4*).

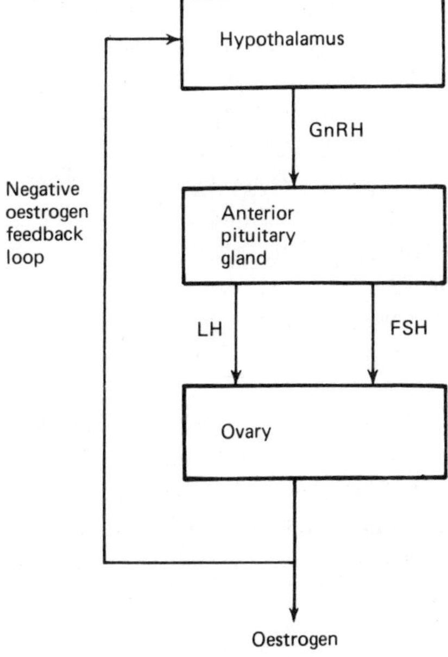

Figure 2.4 An example of a negative feedback loop

Indeed, most feedback loops are negative, and hence regulatory (such as the one described above), although a few positive feedback loops do occur. In addition to being either positive or negative, feedback loops also fall into two further categories, these being known as 'long' and 'short' loops. 'Short' feedback loops are those of the pituitary hormones to the hypothalamus and pituitary gland, whereas the 'long' loops generally refer to the feedback of gonadal steroids to these same controlling centres.

The hormone LH gives an example of a short negative feedback loop. The circulating level of this hormone is monitored at the hypothalamus so that higher blood levels of LH would tend to lower the release of GnRH, and vice versa. Thus LH has a negative short feedback loop on its own production and therefore exhibits controlled release from the pituitary gland. In contrast, in the female, FSH, the other pituitary gonadotrophic hormone, is thought to have a positive short feedback loop. This results in higher circulating levels of FSH stimulating further pituitary release of FSH. However, it should be noted at this point that the short feedback loops are only considered to be of secondary importance in the control of reproductive function.

On the other hand, the long feedback loops are considered to exert the principal controlling influence on reproductive function. The most important of these long loops are those of the gonadal hormones, oestrogen, progesterone and testosterone, to the hypothalamus. All three gonadal steroids have negative feedback loops to the hypothalamus. These negative loops result in a restriction of the output of GnRH from the hypothalamus and, hence, a restriction on the release of the gonadotrophic hormones from the pituitary gland.

However, in addition to its negative loop, oestrogen also forms a positive feedback loop to the hypothalamus under certain conditions. This loop feeds back to the pre-optic region of the hypothalamus (a mechanism which is discussed in detail in Chapter 3) where it stimulates the release of GnRH and, therefore, gonadotrophins.

Two further feedback loops must also be mentioned at this stage. The first of these is that of prostaglandin from the uterus to the ovary. Prostaglandin is produced by the uterus and forms part of a mechanism whereby the uterus informs the ovary, and hence the hypothalamus via the ovarian hormones, that conception has not occurred following ovulation. The prostaglandin is released into the uterine vein of the non-pregnant animal at about day 16 after ovulation, although it should be noted that this release is blocked if a pregnancy has been established. The prostaglandin is then transferred from the uterine vein to the ovarian artery. Once at the ovary the prostaglandin induces regression of the corpora lutea, and hence allows the next batch of follicles to grow and develop.

The final feedback loop to be discussed at this stage is that controlling pituitary hormone release during lactation. This loop is neural, rather than endocrinal, and is initiated by the suckling stimulus. The suckling stimulus sets off a neural reflex which produces an inhibitory influence on the release of GnRH from the hypothalamus. Thus, during lactation, the suckling stimulus results in an inhibition of the release of gonadotrophic hormones, and hence a blockage of reproductive function. Furthermore, since PIH is also inhibited by the suckling stimulus, more prolactin is released into the circulation and thus milk synthesis is stimulated.

All the preceding information may now be summarized by including the feedback loops into our diagrammatic models of reproduction. This has been done in *Figure 2.5*. Using this model, it is now possible to move on to consider each section of the pig's reproductive life cycle individually, bearing in mind that we need to know not

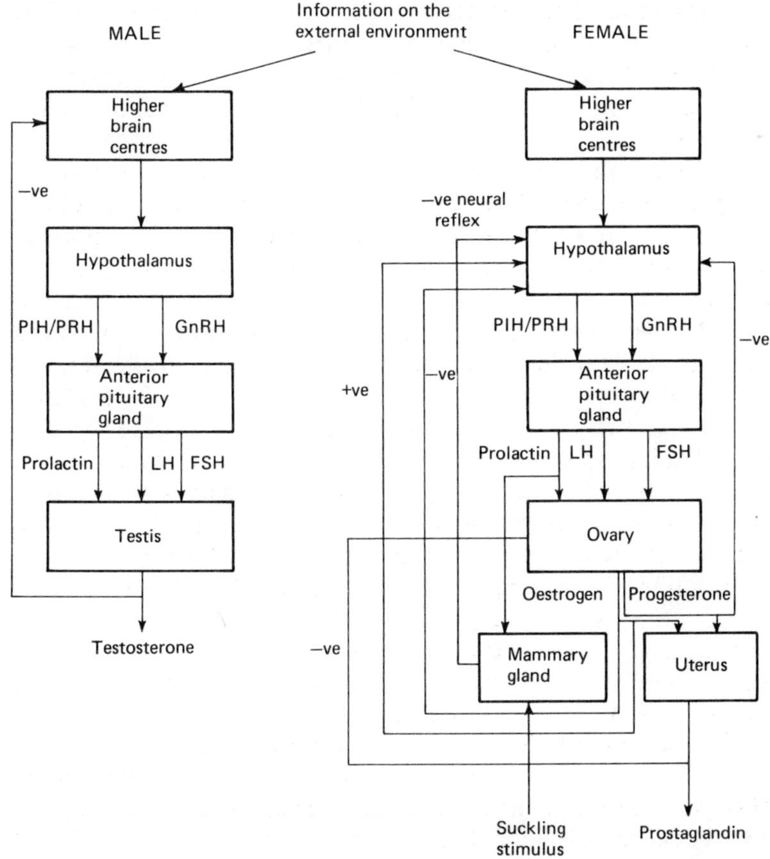

Figure 2.5 A diagrammatic representation of the tissues, hormones and feedback loops involved in reproduction (excluding the short feedback loops)

only how it is naturally controlled but also how it may be controlled and/or stimulated for commercial purposes. The information that has been dealt with so far is, of course, only a simplified model of the whole controlling and controlled systems in reproduction. However, this basic knowledge may now be used, and expanded upon, in each of the subsequent chapters to explain the controlling mechanisms involved at each stage of reproduction.

Part one

The female

Chapter 3

Puberty in the gilt

Puberty may be defined as that phase which links immaturity and maturity, recognized in the gilt by the occurrence of the first oestrous period. This first, or pubertal, oestrus is almost invariably fertile and hence represents the beginning of reproductive capability in the gilt. Puberty normally occurs at about 200 days, although it has been observed to occur in gilts as young as 135 days and as old as 250 days and above. Such variations in age at puberty are due to stimulatory and inhibitory influences originating in both the external and internal environments. These influences include such factors as the breed and genotype of the gilt, nutritional status and the climatic environment. In practice it is possible to manipulate these influences such that gilt age at puberty is reduced. This has the obvious advantage of reducing the cost of rearing replacement gilts since the non-productive period from birth to first mating is shortened. However, the cost and ease of application of any system of puberty stimulation must be borne in mind, as must the effectiveness of the stimulus and its influence on subsequent reproductive performance.

Before such alterations to the normal rate of maturation can be discussed, it is pertinent to consider the endocrine changes occurring during the prepubertal period.

3.1 The physiology of puberty

First a note is required on the origins of much of the information used in this section. The majority of experiments carried out to elucidate the steps in the maturation process have used laboratory animals and although much of this work is now being extended to include commercial species, this information is, at present, limited. Extrapolation from results obtained on laboratory species to commercial animals might be considered unwise, but the more recent experimental results relating to the pig do indicate that the mechanisms involved are similar, and therefore the review is both valid and worthwhile.

Furthermore, this discussion is considered necessary as the mechanisms described will be used in the later sections of the chapter to explain the various systems of puberty stimulation that are available.

This section therefore deals with the mechanisms involved in the normal genetic determination of the onset of puberty, and also the alteration of the normal sequence by the particular environment in which the animal is reared. To accomplish this aim the experimental evidence relating to the endocrinology of this period is reviewed. Finally, an integrative model of the mechanism of the onset of puberty is offered as a working hypothesis.

THE GONADOTROPHINS

The two gonadotrophic hormones, follicle stimulating hormone (FSH) and luteinizing hormone (LH), are primarily responsible for

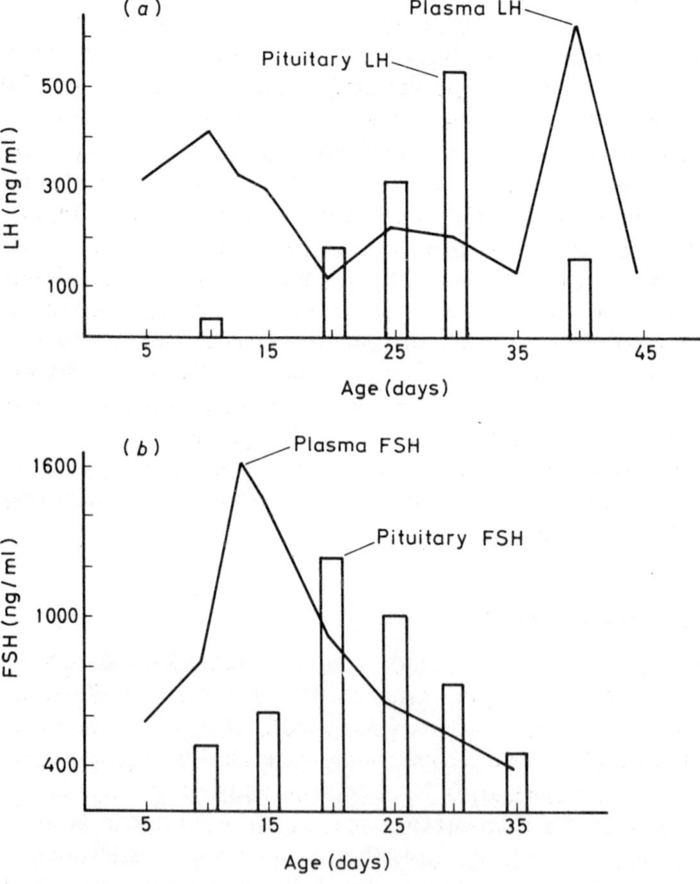

Figure 3.1 Pituitary and plasma LH and FSH levels as a function of age in the immature female rat (from Ramirez, 1973)

the maturation of follicles and their subsequent release from the ovary as mature ova at the time of ovulation. It is therefore clear that the rate of synthesis and release of these hormones by the anterior pituitary gland will have an important effect on the time at which first ovulation (i.e. puberty) occurs.

The levels of these two hormones in the plasma and pituitaries of prepubertal female rats are shown in *Figure 3.1*. For LH it appears that there is a high rate of release in the very young animal, but that this is later inhibited during the period immediately prior to puberty attainment. Thus, soon after birth circulating levels of LH are high but pituitary content is low, whereas, in the later prepubertal stage, release of the hormone into the circulation is reduced resulting in higher concentrations in the pituitary gland. Finally, at puberty an ovulatory surge of LH is seen. This may be interpreted as meaning that the synthesis of LH is maintained throughout the prepubertal period, but that its release is curtailed by an inhibitory influence which becomes effective at a time between birth and puberty.

A similar pattern of change is seen for FSH (*Figure 3.1*(*b*)). This hormone again demonstrates an initial postnatal rise in circulating levels, this being followed by a period of inhibition of its release. However, for FSH it appears that this inhibition is also effective at the level of synthesis of the hormone, since both plasma and pituitary levels of FSH fall in the later prepubertal period. Also, it is considered likely that pituitary FSH content does increase immediately before puberty, this store of the hormone being largely released at puberty (Corbin and Daniels, 1967; Parlow, Anderson and Melampy, 1964).

To sum up, it appears that the prepubertal period may be divided into two separate phases. The first is characterized by high levels of circulating gonadotrophins and low pituitary levels, indicating that no controlling system has evolved to limit the release of these hormones. In contrast, the second phase appears to be the result of an inhibition of the release, and possibly the synthesis, of LH and a definite inhibition of both the synthesis and release of FSH.

THE HYPOTHALAMIC–PITUITARY UNIT

The described changes in the synthesis and release of gonadotrophins from the pituitary gland imply changes in the synthesis and/or release of the hypothalamic releasing hormones (*see* Chapter 2). However, the fact that FSH and LH are being released into the circulation in the early postnatal period would appear to indicate that the releasing hormone (GnRH) is present very early in life. Furthermore, the portal blood system is sufficiently developed to transport this hormone to the pituitary gland which, in turn, is responsive to them. Thus, it appears unlikely that sexual development in the young female is delayed by either the synthesis or release of GnRH, or the ability of the pituitary to respond to this factor.

THE OVARY

During the period from birth to puberty considerable growth and differentiation of the ovary occurs. Furthermore, it appears that ovarian growth is not uniform at this time: more follicles grow during the early prepubertal period and these follicles develop faster than in the older immature female (Pedersen, 1969; Desjardins and Hafs, 1968). This may in part be due to the higher circulating levels of gonadotrophins in the early postnatal period. However, further evidence also suggests that ovarian responsiveness to gonadotrophins varies during the prepubertal period (Price and Ortiz, 1944; Zarrow, Clark and Denenberg, 1969).

The actual pattern of ovarian development varies considerably between species. For example, Graafian follicles have been observed at birth in the heifer, whereas vesicular follicles do not appear until at least 70 days after birth in the gilt (Casida, 1935; Casida *et al.*, 1943). However, irrespective of species, a full set of Graafian follicles are produced by the prepubertal female in readiness for ovulation at puberty. At this stage it has been observed that in many species, including the pig, 'waves' of follicles grow and recede (Hollandbeck *et al.*, 1956). This is presumably due to endocrine conditions being unfavourable for ovulation at this time, and is a theme we shall return to later in the chapter.

OVARIAN HORMONES

The primary hormone produced by the ovary of the prepubertal female is oestrogen, released from the ovarian follicles in response to gonadotrophic stimulation. As mentioned earlier, ovarian responsiveness to the gonadotrophins is variable, maximum responsiveness occurring at about midway through the prepubertal period. This may explain the low levels of circulating oestrogen observed in the early postnatal period (Goldman *et al.*, 1971; Kelch *et al.*, 1972; Presl *et al.*, 1969), since the high blood levels of gonadotrophins would result in the secretion of considerable quantities of oestrogen if the ovary was responsive at this time. As the female approaches puberty oestrogen levels rise steadily, and, during the period immediately preceding puberty attainment there is a major increment in oestrogen secretion (Kelch, 1974).

THE FEEDBACK LOOPS

At this stage we have briefly considered the growth and development of the hypothalamus, anterior pituitary gland and ovary, as related to the production and release of hypothalamic releasing hormones, gonadotrophic hormones and oestrogen. From the discussion of the interrelationships between these hormones it is clear that feedback loops must be present to moderate their synthesis and secretion. It is known that oestrogen forms a negative feedback loop with the hypothalamus to limit the production of hypothalamic releasing hormones, and hence, pituitary gonadotrophins. This feedback loop is known to be operative in both the adult and the late prepubertal female, but not in the early postnatal period (Baker and Kragt, 1969; De Rubin and Rubenstein, 1966). Thus, the absence of the negative oestrogen feedback loop in very young animals explains the high blood levels of gonadotrophins found at this time.

Indeed, it is now clear that the development of a functional negative oestrogen feedback loop is dependent on the development of oestrogen-binding receptors at both the hypothalamic and pituitary level, which first appear midway through the prepubertal period (Alvarez and Ramirez, 1970; Kato, Atsumi and Inuba, 1974). Once these binding sites have developed there is a marked reduction in the circulating levels of both FSH and LH (*Figure 3.1*).

The level of inhibition achieved by a given dose of oestrogen also varies with the maturation of the female. It has been found that the amount of oestrogen necessary to moderate gonadotrophin synthesis and secretion in the young animal is considerably less than that

Table 3.1. Relative amounts of oestrogen necessary to control adult/infantile gonadotrophin system

Conditions	*Amount of oestrogen, infantile/ adult*	*Authors*
Castration cells: effect of oestrogen (rat)	0.2	Schoeller *et al.* (1936)
Gonadotrophins in parabiotic rats, and oestrogen	0.2	Byrnes and Meyer (1951)
Castration cells: effect of oestrogen (rat)	0.20	Hoogstra and De Jongh (1955)
LH in castrated female rats, and oestrogen.	0.33	Ramirez and McCann (1963)

From Ramirez (1973).

required by the adult (*Table 3.1*). The hypothalamic structure responsible for monitoring ovarian steroid levels is termed the 'gonadostat', and the change of sensitivity in the later prepubertal period is referred to as a change in the 'operating level of the gonadostat'. This change in operating level allows the previously repressed gonadotrophin, and hence oestrogen, levels to rise. However, this is not the final step in the maturation process.

It has been known for some time that the administration of oestrogen to either late prepubertal rats (Coppola and Perrine, 1965; Docke and Dorner, 1965; Hagino, Ramaley and Gorski, 1966; Kannwischer, Wagner and Critchlow, 1967; Motta *et al.*, 1968; Smith and Davidson, 1968) or gilts (Hughes, 1976; Hughes and Cole, 1978), can induce premature ovulation. Since under the system so far outlined these higher levels of oestrogen would cause decreased rather than increased levels of gonadotrophins, this observation requires further explanation. It would appear that at this stage of development a positive feedback loop becomes operative, oestrogen now causing increased levels of gonadotrophin release (Davidson, 1969; Smith and Davidson, 1968).

It has been demonstrated that at this time oestrogen feeds back to two separate areas of the hypothalamus, the medium eminence being the centre for negative feedback. The higher levels of oestrogen, resulting from the shift in the gonadostat, feedback to the pre-optic area of the hypothalamus (i.e. the positive feedback area) while at the same time sensitizing the pituitary to the action of the releasing hormone.

Clearly, once this system is operative, higher levels of oestrogen will result in increased release of gonadotrophins, which in turn will result in increased release of oestrogen. This is a self-accelerating cycle which will quickly result in an ovulatory surge of gonadotrophins and hence the attainment of puberty.

HIGHER BRAIN CENTRES

From the discussion so far it appears that certain intrahypothalamic areas are concerned with the maturation process. Thus, for example, we may stimulate the early attainment of puberty by destruction of the negative feedback area of the hypothalamus, thus causing higher levels of circulating gonadotrophins and hence early maturity.

It is also apparent that some extrahypothalamic structures are concerned in the initiation of puberty. Prominent among these structures is the limbic system, notably the amygdaloid complex and the hippocampus. These appear to be mainly involved in the transmission of maturation-inhibiting influences to the hypothalamus along neural pathways such as the stria terminalis (an amygdalo-hypothalamic tract). Since little is known about most higher brain centres, it is of little value discussing their influence on puberty attainment in depth. However, it may be postulated that both excitory and inhibitory extrahypothalamic influences are present, and that these influences produce their effects through changes in the activity of the hypothalamic region.

A MODEL OF PUBERTY ATTAINMENT

A model that describes the interrelationships between the main components of the gonadotrophin control system and its possible sequence of maturation is presented below. It should be emphasized that such a model is partly theoretical, although it does embrace the known facts about puberty attainment in the female.

Description of the model

The components of the system and the variations in their interdependent operations as a function of development are shown in *Figure 3.2*. The controlling system is formed by the anterior pituitary gland, hypothalamus and higher brain centres, the controlled system being the gonads. Ovarian stimulation is provided by FSH and LH, this resulting in the secretion of oestrogens. The system is controlled by the feedback loops of oestrogen and the gonadotrophins, by long and short loops respectively, to the hypothalamic unit of the controlling system. Ovarian function is dependent on the manipulating variables and changes in gonadal properties (capacity to synthesize sex steroids and to respond to the gonadotrophins).

The higher brain centres represent the pineal gland, olfactory lobes, and all other central nervous system centres which monitor environmental changes and relay these to the hypothalamic controlling system. No specific inputs or outputs are given for these centres on the model since it is this component which will be primarily responsible for those changes in maturation due to that particular environment in which the animal is reared. Thus, such environmental factors as male presence, stress, photoperiod change and, possibly, nutrition will be monitored here, their effects being produced at the hypothalamic level via the relay of information from these centres to the gonadotrophin controlling system.

Maturation and role of the components

The normal events that lead to puberty, and the physiological changes described above, allow the division of the maturation process into the following phases in the female:

(1) maturation of the positive feedback action of FSH (infantile phase);
(2) maturation of the negative feedback action of oestrogen (early prepubertal phase);
(3) change in the gonadostat (mid-prepubertal phase);
(4) maturation of the positive oestrogen feedback loop (late prepubertal phase);
(5) LH release and ovulation (ovulatory phase).

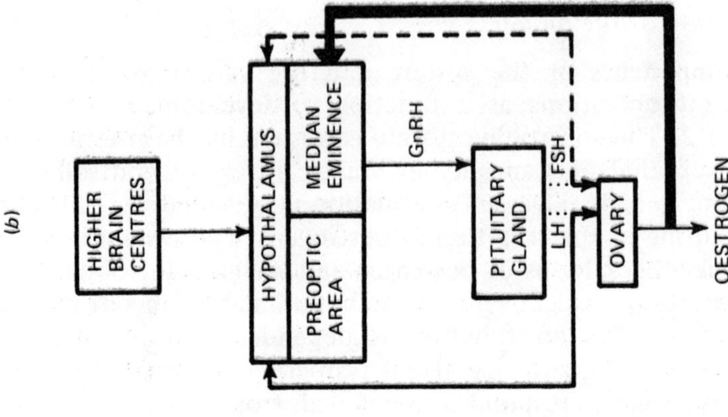
(b)
HIGHER BRAIN CENTRES
HYPOTHALAMUS
PREOPTIC AREA
MEDIAN EMINENCE
GnRH
PITUITARY GLAND
LH
FSH
OVARY
OESTROGEN

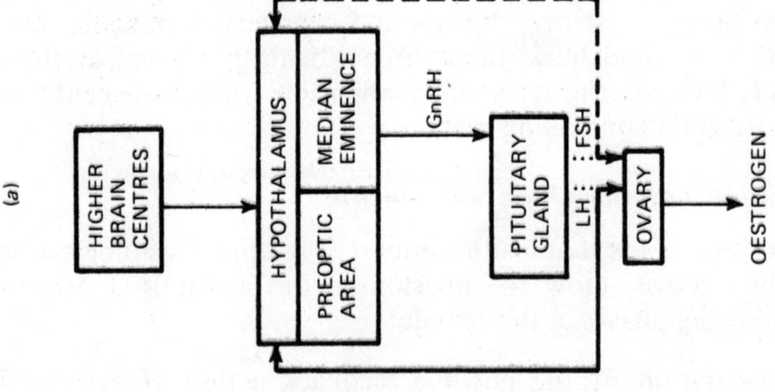
(a)
HIGHER BRAIN CENTRES
HYPOTHALAMUS
PREOPTIC AREA
MEDIAN EMINENCE
GnRH
PITUITARY GLAND
LH
FSH
OVARY
OESTROGEN

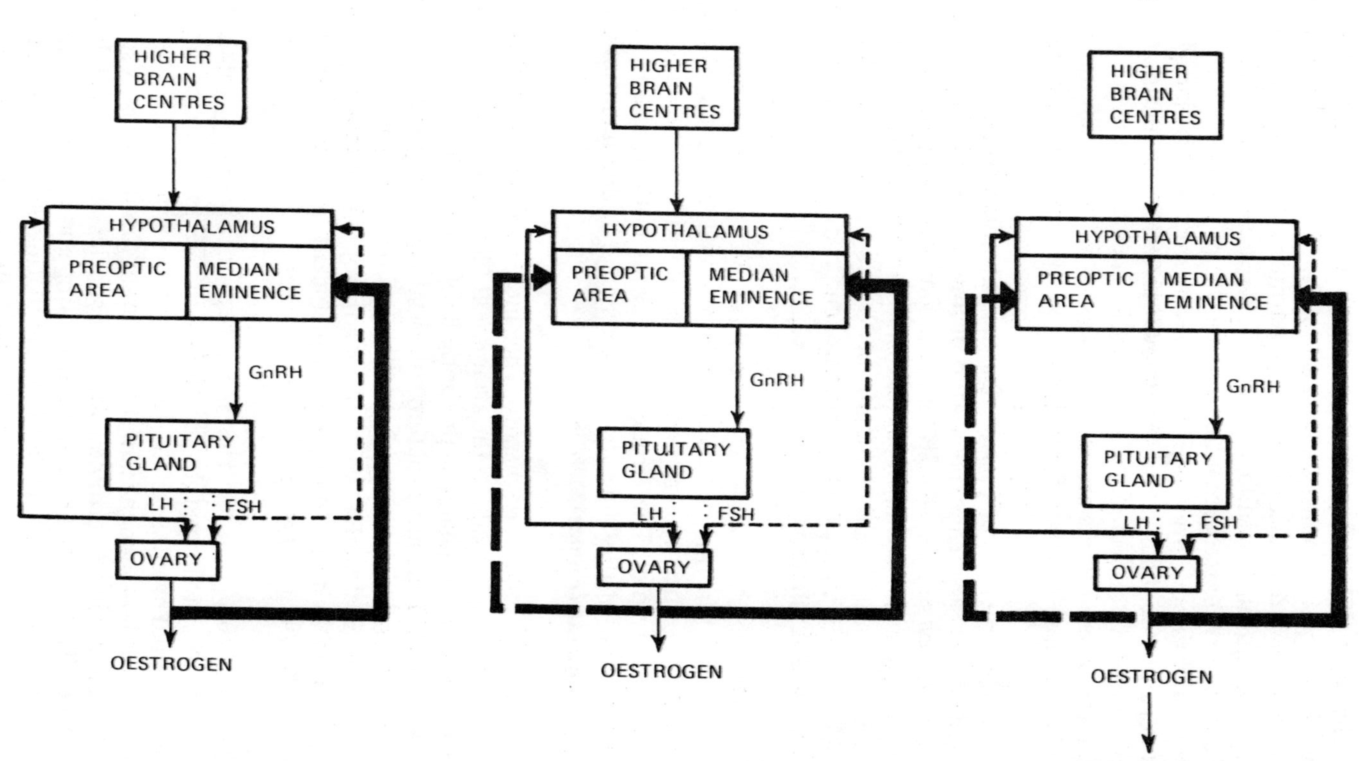

Figure 3.2 A theoretical model of puberty attainment in the gilt (from Hughes, 1976). —— negative feedback loop; – – – – positive feedback loop

(1) *Infantile phase* This phase is characterized by unusually high blood levels of FSH, the evidence indicating that FSH may have a positive feedback action on its own secretion. Therefore, circulating FSH feeds back into the controlling system inducing further pituitary FSH release, through modifications of GnRH metabolism. Also, the suggested lack of oestrogen action during this phase may explain the relatively high blood levels of LH. However, since LH is thought to have a negative feedback action on its own secretion circulating levels of FSH would be expected to be relatively greater than those of LH at this time. These effects are shown in *Figure 3.2*(*a*), which indicates that only the short loops are feeding back into the controlling system.

(2) *The early prepubertal phase* During this phase (*Figure 3.2*(*b*)) the negative oestrogen feedback loop becomes operational, decreasing the pituitary release of FSH and LH. This occurs when the hypothalamic oestrogen binding receptors first become active. The interplay between the negative action of oestrogen and the positive feedback of FSH may explain the exponential decline in circulating FSH observed at this time.

(3) *The mid-prepubertal phase* This third phase supposes that the reference level for the negative oestrogen feedback loop changes, such that the control system becomes less sensitive to the inhibitory effect of the steroid. The circulating levels of both gonadotrophins and steroids rise at this time, due to the change in the operating level of the gonadostat (*Figure 3.2*(*c*)).

(4) *The late prepubertal phase* The increased concentration of circulating oestrogen leads to the attainment of an operating threshold at the anterior hypothalamus–pre-optic level. Once this threshold has been achieved, a positive action of oestrogen on gonadotrophin secretion becomes operative. Once this positive oestrogen feedback loop has been established levels of gonadotrophins increase dramatically, resulting in the production of sufficient oestrogen to trigger ovulation (*Figure 3.2*(*d*)).

(5) *The ovulatory phase* The main feature of this phase is the release of LH, facilitated by oestrogen in the presence of progesterone. The final event after the LH surge is ovulation.

The maturation sequence of the different components described above is genetically determined for each species and for each individual. Once the first two phases are operating and the ovary is able to respond to changes in the level of circulating gonadotrophins the function of the system can be advanced by different stimuli, and precocious puberty may be initiated. The animal is particularly responsive to such external stimulation during the late prepubertal phase, at which time the system is primed for puberty attainment.

3.2 Factors affecting the attainment of puberty

It has been shown above that the onset of puberty is genetically determined and that the course of events culminating in the attainment of puberty may be altered by environmental factors. The purpose of this section is to discuss those factors which influence the attainment of puberty in the gilt, and to relate this information to the theoretical model given in *Figure 3.2.*

AGE, LIVEWEIGHT AND RATE OF GROWTH

These three factors are intimately interrelated, and also have a close relationship with nutrition. They are considered together here since their individual effects on puberty attainment are difficult to assess because of these interrelationships.

It has been reported that the attainment of puberty in the gilt is more a function of age than weight, and normally occurs at about 200 days (Duncan and Lodge, 1960). However, many reports suggest that considerable variation exists, with puberty being attained as early as 135 days (Hughes and Cole, 1975) and as late as 251 days (Warnick *et al.*, 1951). This is hardly surprising when one considers the multitude of contributing factors influencing puberty attainment.

Indeed, since age at puberty will vary according to the genetic potential of the gilt and the environment in which it is reared, it would seem reasonable to consider other measurements of maturation. Of the alternatives physiological age, or the degree of maturity of the hypothalamic–pituitary–ovarian axis, would appear to be the most appropriate. Thus, variations in physiological age, in gilts of similar chronological ages, may be used to explain variations in response to external stimulation.

Although chronological age is undoubtedly an inaccurate measure of the degree of maturity of a gilt, it seems that it is a significantly better measure than liveweight (Robertson *et al.*, 1951a, b). Thus, although chronological age does not provide a definitive method of determining stage of maturation, it does appear to be more closely associated with physiological age than is the body weight of the animal. It should be added that in this respect the pig appears to differ from many other species, including the human (Frisch and Revelle, 1970) and the mouse (Monteiro and Falconer, 1966). However, despite the relative unimportance of live weight to the maturation process, it is logical to assume that a lower limit of body weight exists below which puberty will not be attained.

When body weight and gilt age are combined we can calculate the growth rate of the animal. This is a factor which is considerably altered by nutritional status, but which is also thought to have an influence *per se* on puberty attainment. Brody (1945) concluded that the attainment of sexual maturity is closely connected with the point of inflection of the growth curve, and that the age but not the weight at which the point of inflection occurs will be altered by the level of early nutrition. This implication, that puberty occurs at a fixed point on the growth curve, has received some support (e.g. Monteiro and Falconer, 1966) but does not appear to be applicable to the gilt (Burger, 1952; Goode, Warnick and Wallace, 1965; Gossett and Sorensen, 1959; Haines, Warnick and Wallace, 1959; Sorensen, Thomas and Gossett, 1961). However, it should be added that a positive genetic relationship does exist between age at puberty and weight gain (Reutzel and Sumption, 1968).

Finally, in an effort to explain growth rate influences on puberty attainment, Phillips and Zeller (1943) have suggested that faster growing animals mature earlier because they have a higher level of pituitary activity. It should, however, be stated that the evidence on this point is far from conclusive.

NUTRITION

Live weight and rate of growth are interrelated with the nutrient supply to the animal and, thus, it might be expected that both plane of nutrition and the composition of the diet would influence puberty attainment in the female.

Plane of nutrition

The experiments which have been carried out to elucidate the relationship between plane of nutrition and puberty attainment have yielded variable and conflicting results. For example, some workers have concluded that a high level of feeding in the prepubertal period results in puberty advancement (Burger, 1952; Haines, Warnick and Wallace, 1959; Lazauskas, 1965; Meyer and Bradford, 1974; Short and Bellows, 1971; Zimmerman *et al.*, 1960), while others have reported no significant differences in puberty attainment associated with variations in plane of nutrition (Christian and Nofziger, 1952; Lodge and MacPherson, 1961; Robertson *et al.*, 1951b). Yet others have reported a delay in sexual maturity following high plane feeding (Self, Grummer and Casida, 1955). These results are summarized in *Table 3.2*. It may be speculated that these conflicting results were, at least in part, due to other variables (such as boar introduction – c.f. Christian and Nofziger, 1952; Pay and Davies, 1973). Furthermore, it

Table 3.2. The effects of plane of nutrition on puberty attainment in the gilt

High plane diet		*Low plane diet*		*Source*
Age (days)	*Weight* (kg)	*Age* (days)	*Weight* (kg)	
178	86.0	176	52.7	Lodge and MacPherson (1961)
223	98.2	208	69.1	Self *et al.* (1955)
170	77.3	167	58.2	Christian and Nofziger (1952)
188	88.2	235	53.6	Burger (1952)
195	89.1	217	72.7	Haines *et al.* (1959)

is probable that these effects are caused by the energy content of the diets rather than level of feed intake *per se*.

Energy intake

The relationship between energy intake and gilt age at puberty exhibits similar variation to that discussed for plane of nutrition and pubertal age. However, it is worth noting that the level of metabolizable energy (ME) intake of pigs on the restricted diets in many of these experiements carried out to elucidate this relationship were adequate for an acceptable rate of growth. This may partly account for the minimal difference in age at puberty frequently observed in these experiments, as, indeed, may differences in breed, age at start of feed restriction and dietary composition.

Many reports suggest that restricting energy intake results in a delay in puberty attainment (Goode, Warnick and Wallace, 1961, 1965; Haines, Warnick and Wallace, 1959; O'Bannon, 1964; Zimmerman *et al.*, 1960), although other evidence suggests that the reverse is true (Etienne and Legault, 1974; Gossett and Sorensen, 1959; Hafez, 1960; O'Bannon *et al.*, 1966; Sorensen, Thomas and Gossett, 1961). This work was summarized by Anderson and Melampy (1972) who observed that restricting energy intake delayed puberty an average of 16 days in nine experiments, whereas the restricted diet hastened the onset of puberty by 11 days in five other experiments (*Table 3.3*).

Table 3.3. The effects of metabolizable energy intake on puberty attainment in the gilt

Number of trials	*ME intake* (MJ/day)		*Pubertal age* (days)		*Pubertal body wt* (kg)	
	Restricted	*Full*	*Restricted*	*Full*	*Restricted*	*Full*
9	23.2	37.5	217	201	74	91
5	25.2	37.5	201	212	74	94

From Anderson and Melampy (1972).

Thus, it may only be concluded that this problem is unresolved, although it does seem that severe restriction of energy intake does tend to delay puberty.

Protein

It appears likely that any effect of protein on the maturation process will only become apparent under conditions of either severe restriction or amino acid imbalance (Angelova, 1965; Duee and Etienne, 1974). However, retardation of maturation has been observed in some experiments where protein has been restricted to a greater or lesser extent (Adams *et al.*, 1960; Baker, 1959; Davidson, 1930). Furthermore, differences in age at puberty caused by the use of either plant or animal protein sources (Fowler and Robertson, 1954) do appear to indicate the involvement of protein quality and, hence, amino acid balance. This is substantiated by the work of Friend (1973) who observed that gilts fed diets supplemented by lysine and methionine attained puberty 12–24 days earlier than control animals.

Fat

The effects of fat level in the diet on sexual development are considered to be minimal under most conditions, since the influence of fat is mainly exerted within the energy component of the diet. However, one report (Witz and Beeson, 1951) does suggest that puberty may be delayed and ovarian growth retarded in gilts fed a fat-free diet from an early age.

Vitamins and minerals

Data on the effects of vitamins and minerals on reproduction in the pig are limited. However, it seems likely that severe deficiencies of vitamin A (Hughes, 1934), vitamin B_{12} (Johnson, Moustgaard and Højgaard-Olsen, 1952) and manganese (Plumlee *et al.*, 1956) may retard sexual development.

GENETICS

It is difficult to assess the effects of genotype on puberty attainment in the gilt, since it requires the standardization of all other stimulatory and inhibitory factors. However, it has been shown that breed differences do exist with respect to age at puberty and that both inbreeding and crossbreeding influence this.

Differences in age at puberty between breeds have, indeed, been shown by many workers (Etienne and Legault, 1974; Phillips and Zeller, 1943; Robertson *et al.*, 1951a; Self, Grummer and Casida,

1955; Warnick *et al.*, 1951). However, some reports have suggested that breed differences are not significant, although this conclusion may be due to the great variability in age at puberty observed within breeds (Haines, Warnick and Wallace, 1959; Mirskaia and Crewe, 1930). Irrespective of breed differences it has been repeatedly demonstrated that the crossing of two separate breeds to produce a crossbred gilt can result in earlier maturation by that offspring compared to purebred animals (Foote *et al.*, 1956; Short, Zimmerman and Sumption, 1963; Squiers, Dickerson and Mayer, 1950; Zimmerman *et al.*, 1960).

It has also been reported (Burger, 1952) that significant between-family differences are apparent for gilt age at puberty. Furthermore, despite evidence that the heritability of pubertal age is low or even zero (Reutzel and Sumption, 1968), sire and dam effects on the attainment of puberty in the progeny have been repeatedly demonstrated (Reddy, Lasly and Mayer, 1958; Hughes and Cole, 1975). Thus, for example, Hughes and Cole (1975) found that, using the offspring from four different Landrace boars, pubertal age varied significantly according to sire (*Table 3.4*).

Table 3.4. The effect of sire on puberty attainment in its female offspring

	Boar			
	A	*B*	*C*	*D*
Mean pubertal age (days)	161.1	180.7	180.8	185.2
Range in age at puberty (days)	135–187	140–225	157–196	145–225
Mean wt. at puberty (kg)	81.1	94.1	90.3	92.8

From Hughes and Cole (1975).

Finally, some insight into the mechanisms which produce crossbreeding influences and between-breed differences in pubertal age may be obtained from the work of Daily *et al.* (1970) in the USA. These workers found that pituitary gland weight was significantly higher, at the same age and weight, for crossbreds than it was for purebreds. Thus, increased pituitary activity may well be responsible for earlier maturity in the crossbred gilt. Furthermore, a similar mechanism has been suggested to be operative to produce between-breed differences (Dailey *et al.*, 1970; Ruiz *et al.*, 1968).

ENVIRONMENTAL INFLUENCES

Climatic environment

It is well documented that seasonal effects exist, influencing reproductive function in the female. Indeed, many workers have reported that spring-born gilts attain puberty earlier than those born in other seasons (Corteel, Signoret and Du Mesnil du Buisson, 1964; Robertson *et al.*, 1951a; Schmidt and Bretschneider, 1954; Sorensen, Thomas and Gossett, 1961). In fact, since seasonal effects are mediated primarily by daylength and temperature it seems reasonable to consider the influence of each of these factors on puberty attainment.

Photoperiod has been reported to influence many aspects of reproduction in the pig (Klotchkov *et al.*, 1971). In particular, most workers have reported a decrease in pubertal age following exposure to extended day length (Hacker, King and Bearss, 1974; King, 1961; Martinat *et al.*, 1970; Surmuhin and Ceremnyh, 1970). This effect of daylength on reproductive function appears to be effected via the pineal gland. Light perceived by the retina regulates the activity of the sympathetic nerves to the pineal gland (Wurtman, 1967). These nerves then release a neurotransmitter which controls the formation of the enzyme 5-hydroxy-indole-*o*-methyl-transferase in the pineal. This enzyme, in turn, controls melatonin synthesis. This synthesis is thought to occur predominantly in the dark, and hence extended daylength could lead to a reduced production of melatonin. Since melatonin is thought to inhibit the synthesis and/or release of gonadotrophins from the pituitary (Wurtman, Axelrod and Chu, 1963), this appears to provide a possible mechanism of action of light on reproductive function. Extended daylength would therefore result in a reduction of the inhibitory influence of melatonin on the production of gonadotrophins.

Many workers have reported that high summer temperatures lead to a reduction in fertility, and hence that moderately cold temperature may enhance reproductive performance (Burger, 1952; Corteel, Signoret and Du Mesnil du Buisson, 1964; Schmidt and Bretschneider, 1954). However, work with rats has shown that extremes of temperature may result in a delay in the attainment of sexual maturity (Donovan and Van der Werff ten Bosh, 1965; Mandl and Zuckerman, 1952). Furthermore, the evidence suggests that temperature effects are mediated by the adrenal cortex, since adrenocorticotrophic hormone depresses rate of growth (Moon, 1937) and cortisone causes a delay in the onset of sexual maturity (Blivaiss *et al.*, 1954).

Thus it seems clear that puberty attainment is affected by both environmental temperature and daylength, which combine to produce seasonal effects. Increasing daylength results in the earlier attainment of puberty whereas increasing temperature appears to delay the onset of sexual maturity. It also seems likely that extremes of either daylength or temperature lead to a delay in puberty attainment.

Social environment

The effects of social factors on reproduction in the pig have received relatively little attention. However, work in laboratory species has indicated that variations in the social environment can have stimulatory (Levine, 1962; Morton, Deneberg and Zarrow, 1963) or, more usually, inhibitory influences (Arvay and Nagy, 1959; Duckett, Varon and Christian, 1963). The only alteration in the social environment which has been found to influence puberty attainment in the gilt is that of transportation. This environmental change has been reported to result in the spontaneous induction of puberty a few days after transport (Du Mesnil du Buisson and Signoret, 1961; Paredis, 1961). However, this treatment only appears to be effective

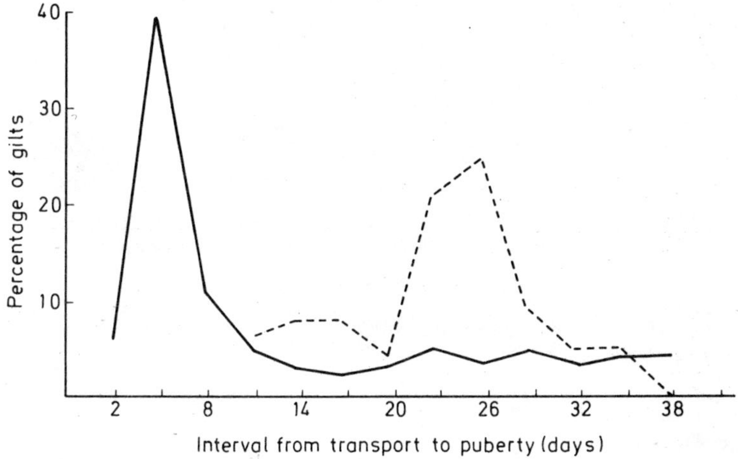

Figure 3.3 The effects of transport on puberty attainment in gilts (from Signoret, 1970). ——— presentation to boar immediate; – – – – – – presentation to boar delayed for 10 days

when the gilt is about the age of puberty. In addition, Signoret (1970) demonstrated that this effect was independent of male stimulation, since when presentation to a boar was delayed by 10 days following transportation the peak of oestrus occurrence was unchanged (*Figure 3.3*).

THE MALE EFFECT

The introduction of a mature boar to immature gilts is known to induce the precocious attainment of puberty. The data from various authors (Brooks and Cole, 1969, 1973; Brooks, Pattinson and Cole, 1970; Zimmerman, Carlson and Nippert, 1969; Hughes and Cole, 1976, 1978) have been combined in *Figures 3.4* and *3.5* to demonstrate the effects of male introduction at various gilt ages on both the interval from first boar contact to puberty and the actual age at puberty itself. This data shows that the introduction of a boar to gilts below about 140–150 days results in an extended period of

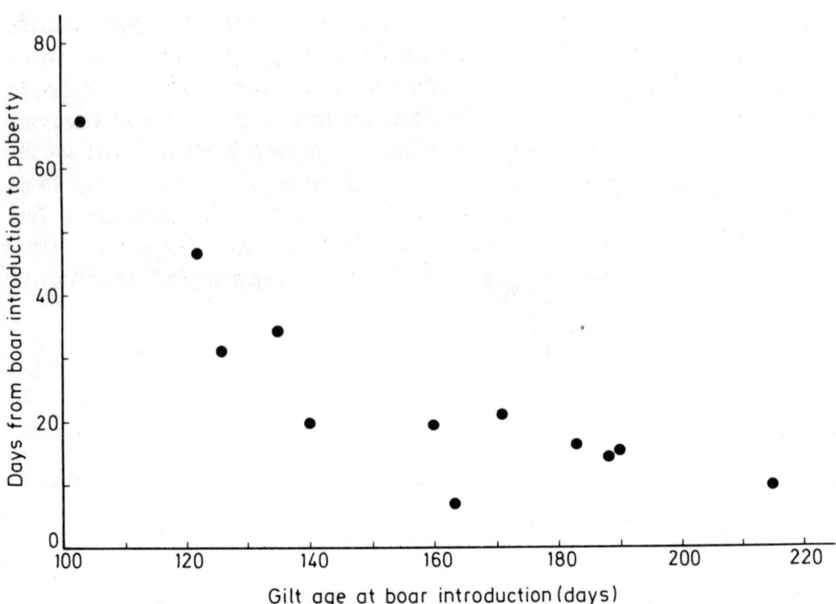

Figure 3.4 The effects of gilt age at boar introduction on the interval from first boar contact to puberty

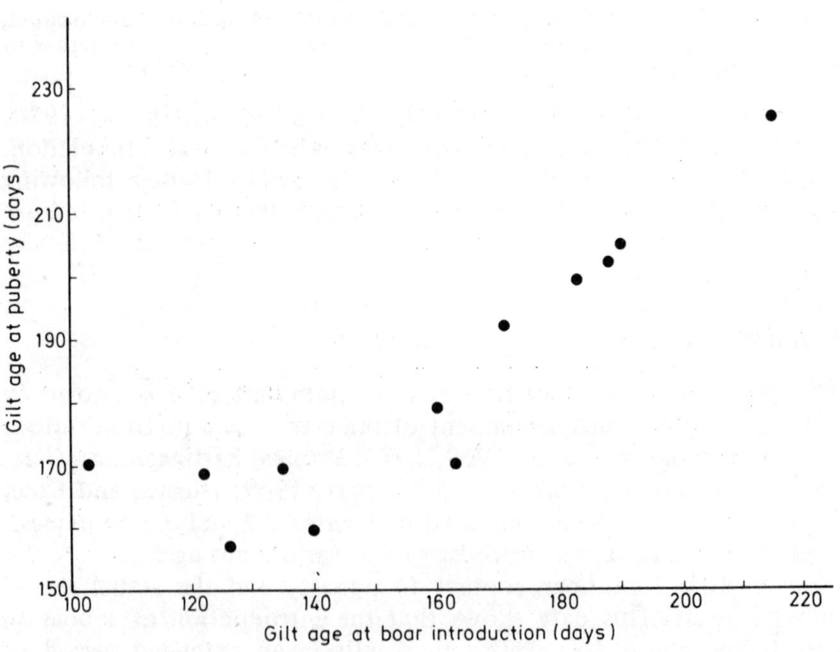

Figure 3.5 The effects of gilt age at boar introduction on age at puberty

male contact prior to puberty attainment, and only a slight reduction in pubertal age. On the other hand, when gilts are first introduced to a boar at 180 days or above, the age at puberty is considerably increased. Thus, optimum gilt age for first boar contact would appear to be within the range 150–170 days. Furthermore, when boar introduction first occurs at this age maximum synchronization of the pubertal oestrus is obtained (Brooks and Cole, 1973; Hughes and Cole, 1976). This effect may, indeed, be further reinforced if a rotation of boars is used (Brooks and Cole, 1970), giving a daily renewal of the stimulus (*Figure 3.6*).

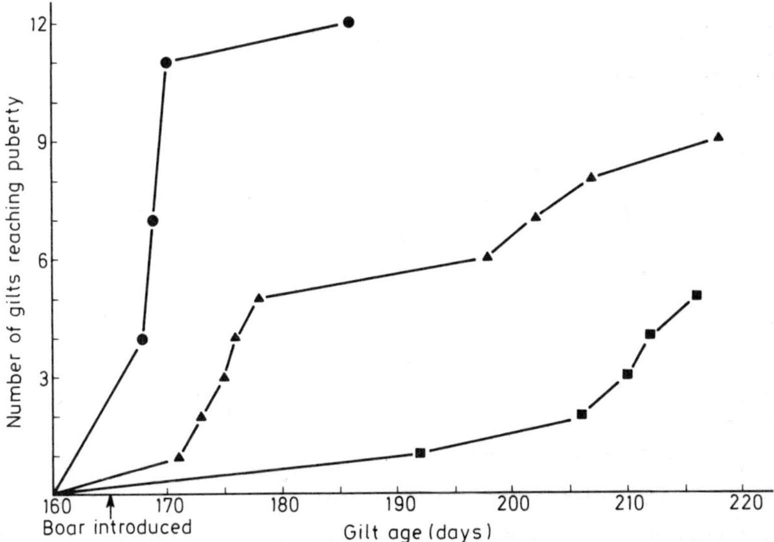

Figure 3.6 The effects of using a single boar or a rotation of boars on oestrus synchronization in the prepubertal gilt (from Brooks and Cole, 1970). ● rotation of boars used; ▲ one boar used; ■ no boars used

The mechanism whereby the male stimulates puberty attainment in the gilt is not fully understood. Experiments in mice (Bronson and Desjardins, 1974; Bronson, 1975) have indicated that, in this species at least, male introduction results in a rapid increase in the circulating level of LH. This is then followed by a dramatic rise in plasma oestrogen levels which lasts for about 52 hours with two distinct peaks being observed during this period (*see Figure 3.7*). Finally, a peak of LH secretion occurs, this resulting in the pubertal ovulation. On the basis of these results, Bronson and Desjardins (1974) concluded that 'male stimuli, pheromonal or otherwise, promote the final maturation of the positive feedback system by enhancing the secretion of ovarian oestrogens'.

Unfortunately, no change in plasma LH levels has been observed in the gilt in response to boar introduction. In addition, recent evidence indicates that the urinary pheromone produced by the boar

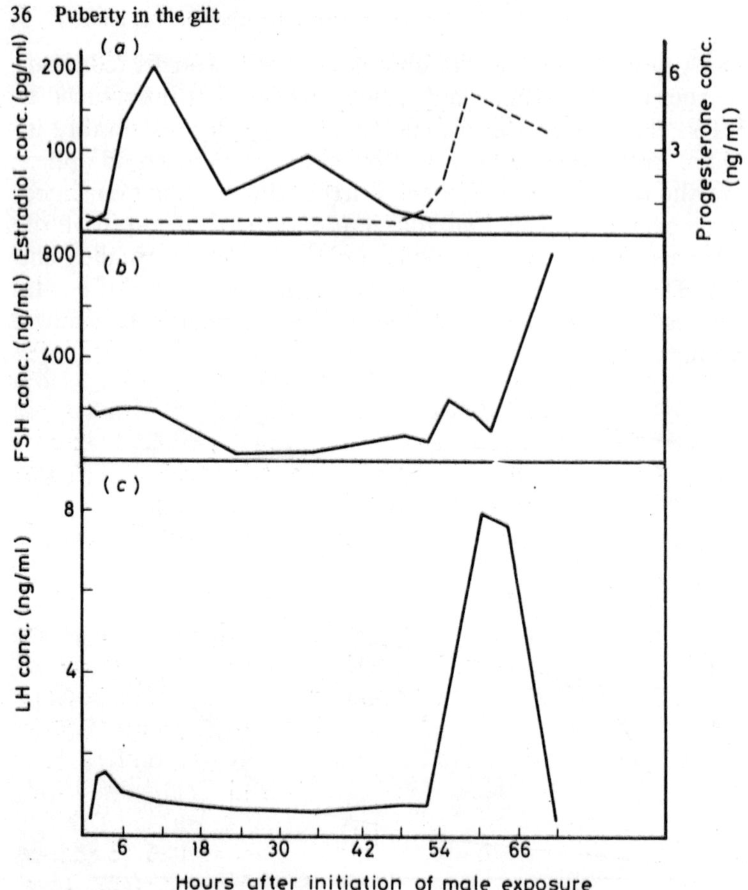

Figure 3.7 Changes in the circulating levels of gonadotrophins and sex steroids following male introduction to prepubertal female mice (from Bronson and Desjardins, 1974). (a) ———— oestrogen; - - - - - - progesterone; (b) FSH; (c) LH

is also ineffective in stimulating precocious puberty in the gilt, although the involvement of a salivary pheromone does seem probable (Kirkwood and Hughes, unpublished data). Therefore, we may only speculate as to the possible mechanism of action of the boar stimulus. It does seem likely that the stimulus provided by the boar is associated with both a stress effect (this probably producing its effects via changes in the rate of prolactin secretion) and the action of a pheromone. However, until such time as more data is available on this subject the source and mechanism of action of the boar stimulus must remain unresolved.

EXOGENOUS HORMONES

Most attempts to induce precocious puberty with exogenous hormones have involved the use of gonadotrophic stimulation. These gonadotrophins have been given in various preparations, the most common of which has been pregnant mare serum (PMS) followed by human chorionic gonadotrophin (HCG). This sytem is used because PMS is known to contain a high concentration of FSH, whereas HCG contains predominantly LH. However, as discussed earlier in the chapter, the effectiveness of any such treatments will be dependent on the age of the gilt at the time of application. It is quite clear that puberty and subsequent cyclicity can only be induced when all the component parts of the hypothalamic–pituitary–ovarian system are approaching maturity and, hence, are able to respond to the stimulus. In some situations it is possible to induce ovulation in the very young female (for example, using a combination of PMS and HCG), but this is unlikely to result in subsequent cyclicity as endogenous hormone release will prove inadequate. In other words, the early prepubertal female may be forced to ovulate by the exogenous administration of the necessary hormones, but as soon as this support is withdrawn the female is likely to revert to a prepubertal state.

In the gilt it appears that ovulation cannot be induced before 100 days and that at this age both ovulation and fertilization rates are low (Casida, 1935; Majerciak *et al.*, 1970). Indeed, the success of exogenous gonadotrophins in the stimulation of true precocious puberty (i.e. a pubertal ovulation followed by either normal pregnancy or subsequent normal cyclicity) has been very limited (c.f. Baker and Coggins, 1968; Dzuik, 1964; Dzuik and Gehlbach, 1966; Schwartz, Robison and Ulberg, 1971). A typical example of this is the work of Dzuik and Gehlbach (1966) who attempted to induce ovulation in gilts of 100–180 days of age using PMS and HCG. They observed that, although 91 per cent of the gilts displayed vulval swelling and 43 per cent exhibited typical oestrous symptoms, only 3 out of 18 gilts were pregnant 23 days after insemination. Furthermore, no young were actually farrowed, and none of the treated gilts showed oestrus 21 days later. Thus, the success rate of PMS/HCG stimulation is low, although it should be added that response to the treatment does improve as the age of the gilts at the time of treatment is increased.

An alternative approach to the induction of early puberty in the gilt is the use of exogenous oestrogen. Such a system is dependent for its success on the maturation of the positive oestrogen feedback loop, by the administered oestrogen. Indeed, the use of exogenous oestrogen has proved effective in inducing oestrus in the prepubertal mouse and rat (Bronson, 1975; Ramirez and Sawyer, 1965), the postpartum ewe (Foote and Matthews, 1969) and the anoestrous gilt (Whetteman, Omtvedt and Williams, 1973). More recently Hughes and Cole (1978) have demonstrated that oestrogen may be used successfully for the induction of precocious puberty in the gilt. Their

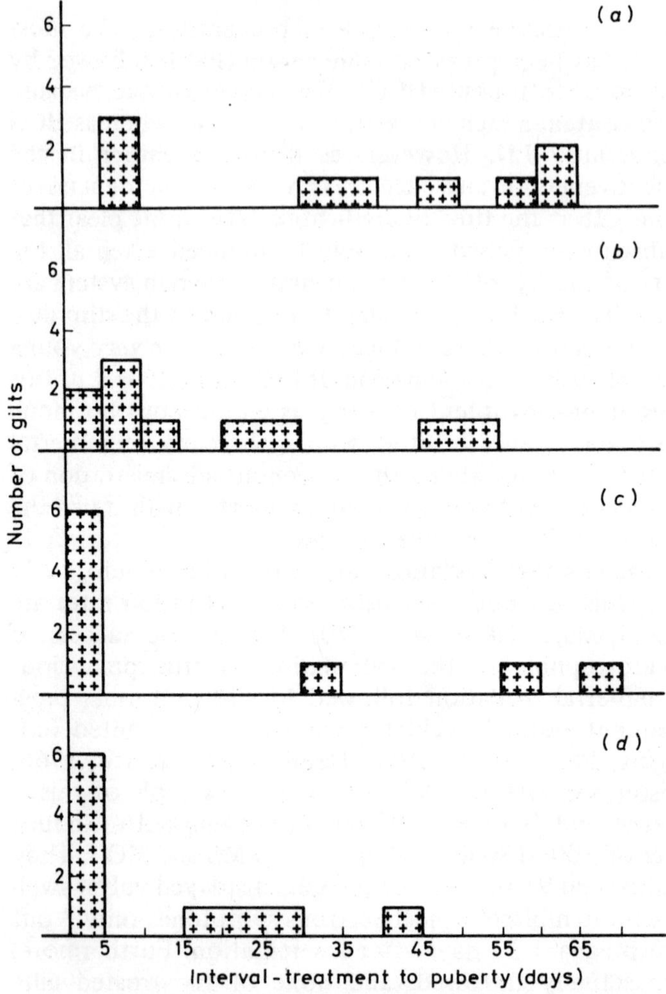

Figure 3.8 The influence of oestrogen treatment and/or boar contact on puberty attainment in the 140-day-old gilt (from Hughes and Cole, 1978). (a) control; (b) control + boar: (c) oestrogen; (d) oestrogen + boar

results show that an early synchronized first oestrus may be precipitated in about 60 per cent of gilts when treatment is given at 140 days (*Figure 3.8*). Furthermore, they speculate that the number of gilts responding to exogenous oestrogen should increase as the age of the gilts, and hence their degree of maturation, is increased.

3.3 Conclusions

In order to summarize the preceding information to provide some guidelines for application to commercial gilt rearing, we must first consider the requirements of the producer. These are:

(1) Early puberty – to reduce rearing costs.
(2) Low cost – the stimulation system costs must not outweigh the advantage of reduced rearing costs.
(3) Ease of application.
(4) Synchronization of oestrus – this provides for easier management and the treatment of gilts as a group.
(5) Normal subsequent reproductive performance – the advantages of early maturity must not be lost in erratic cyclicity after puberty, poorer conception rates at mating or smaller litters at first farrowing.

Three forms of puberty stimulation are possible to achieve these requirements, namely boar contact, the use of exogenous hormones and transportation.

The aim here is to consider the potential of each of these stimulation systems, bearing in mind the needs of the commercial producer. For all three systems it is assumed that the gilt should be provided with a supply of nutrients which is sufficient to support an adequate rate of growth, and an environment which does not limit performance.

THE BOAR EFFECT

The influence of the male has already been described in earlier sections of this chapter. Here it is only necessary to consider at what age a gilt should first be exposed to the boar. Firstly, it must be stated that the cost and ease of application of this system will rise as the number of days from first boar contact to puberty attainment increases. This interval is related to gilt age at boar introduction, the older the gilt at first contact the shorter the interval to puberty (*Figure 3.4*). However, since our objective is to stimulate early puberty, a balance must be found between the length of the interval from first boar contact to puberty and the actual age of the gilt at puberty. Considering the data presented in *Figures 3.4* and *3.5* it appears that the optimum response will be obtained when first boar contact occurs at about 150–160 days of age. This is also the time when maximum synchronization of the pubertal oestrus might be expected to occur. Finally, it is also suggested that different boars be used each day if possible, thus providing a daily renewal of the stimulus.

EXOGENOUS HORMONES

It is concluded above that gonadotrophic stimulation is of limited value in the stimulation of early puberty, but that the use of exogenous oestrogen may have commercial application. However, since this system is still being evaluated, we may only speculate on its final value under commercial conditions. The present evidence suggests that application of this system to gilts of about 150 days should result in an early, synchronized pubertal oestrus. Furthermore, the system is inexpensive and extremely easy to apply. The only doubts at present surrounding this system are its effects on cyclicity after the pubertal oestrus, although it seems likely that these may be overcome by the concurrent employment of daily boar contact.

TRANSPORTATION

This system (c.f. *Figure 3.3*) is only of importance where replacement gilts are bought-in rather than home-reared. Under these conditions, gilts transported to a farm may be expected to attain puberty soon after transportation. However, it should be added that this system only appears to be effective in gilts of 180 days and above.

In conclusion it appears that three systems of puberty stimulation are available. Home-bred gilts may be induced to attain puberty at above 150 days using either boar contact, exogenous oestrogen or a combination of the two. Bought-in gilts, on the other hand, should attain puberty soon after arrival, if transportation occurs at 180 days or above. However, in either case, the growing period must be considered, and adequate care taken over the provision of sufficient food and a suitable environment. Finally, the effects of early maturity on subsequent reproductive performance should be considered. This subject will be dealt with in detail in subsequent chapters, but it is worth noting at this stage that the induction of precocious puberty does not (with the possible exception of exogenous oestrogen treatment) adversely affect subsequent reproductive performance in the gilt.

3.4 References

ADAMS, C.R., BECKER, D.E., TERRILL, S.W., NORTON, H.W. and JENSEN, A.H. (1960). *J. Anim. Sci.* **19**, 1245 (abstract)

ALVAREZ, E.O. and RAMIREZ, V.D. (1970). *Neuroendocrinology* **6**, 349

ANDERSON, L.L. and MELAMPY, R.M. (1967). In *Reproduction in the female mammal* (Ed. by G.E. Lamming and E.C. Amoroso). London, Butterworths

ANDERSON, L.L. and MELAMPY, R.M. (1972). In *Pig Production* (Ed. by D.J.A. Cole). London, Butterworths

ANGELOVA, L. (1965). *Zhivot. Nauk.* **2**, 489

ARVAY, A. and NAGY, T. (1959). *Acta. neuroveg.* **20**, 57

BAKER, B. (1959). *J. Anim. Sci.* **18**, 1160 (abstract)

BAKER, F.D. and KRAGT, C.L. (1969). *Endocrinology* **85**, 522

BAKER, R.D. and COGGINS, E.G. (1968). *J. Anim. Sci.* **27**, 1607

BLIVAISS, B.B., HANSON, R.O., ROSENZWEIG, R.E. and McNIEL, K. (1954). *Proc. Soc. exp. Biol. Med.* **86**, 678

BRODY, S. (1945). *Bioenergetics and growth*. New York, Reinhold

BRONSON, F.H. (1975). *Endocrinology* **96**, 511
BRONSON, F.H. and DESJARDINS, C. (1974). *Endocrinology* **94**, 1658
BROOKS, P.H. and COLE, D.J.A. (1969). *Rep. Sch. Agric. Univ. Nott.* p. 74
BROOKS, P.H. and COLE, D.J.A. (1970). *J. Reprod. Fert.* **23**, 435
BROOKS, P.H. and COLE, D.J.A. (1973). *Anim. Prod.* **17**, 305
BROOKS, P.H., PATTINSON, M.A. and COLE, D.J.A. (1970). *Rep. Sch. Agric. Univ. Nott.* p.65
BURGER, J.P. (1952). *Onderstepoort J. vet. Sci. Anim. Ind. Suppl.* 2
BYRNES, W.W. and MEYER, R.K. (1951). *Endocrinology* **48**, 133
CASIDA, L.E. (1935). *Anat. Rec.* **61**, 389
CASIDA, L.E., MEYER, R.K., McSHAN, W.H. and WISNICKY, W. (1943). *Am. J. vet. Res.* **4**, 76
CHRISTIAN, R.E. and NOFZIGER, J.C. (1952). *J. Anim. Sci.* **11**, 789 (abstract)
COPPOLA, J.A. and PERRINE, J.W. (1965). *Endocrinology* **76**, 865
CORBIN, A., and DANIELS, E.L. (1967). *Neuroendocrinology* **2**, 394
CORTEEL, J.M., SIGNORET, J.P. and DU MESNIL DU BUISSON, F. (1964). *Proc. 5th int. Congr. Anim. Reprod. A.I. (Trento)*, Vol. 3, 536
DAILEY, R.A. CLOUD, J.G., FIRST, N.L., CHAPMAN, A.B. and CASIDA, L.E. (1970). *J. Anim. Sci.* **31**, 937
DAVIDSON, H.R. (1930). *J. agric. Sci., Camb.* **20**, 233
DAVIDSON, J.M. (1969). In *Frontiers of neuroendocrinology* (Ed. by W.F. Ganong and L. Martini). New York, Oxford University Press
DE RUBIN, Z.S. and RUBINSTEIN, L. (1966). *Revta Soc. argent. Biol.* **42**, 43
DESJARDINS, C. and HAFS, H.D. (1968). *J. Anim. Sci.* **27**, 472
DOCKE, F. and DORNER, G. (1965). *J. Endocr.* **33**, 491
DONOVAN, B.T. and VAN DER WERFF TEN BOSH, J.J. (1965). *Physiology of Puberty*. London, Arnold
DUCKETT, G.E., VARON, H.H. and CHRISTIAN, J.J. (1963). *Endocrinology* **72**, 403
DUEE, P.H. and ETIENNE, M. (1974). *Journées de la recherche porcine en France*, pp. 43–47. Paris, L'Institut Technique du Porc
DU MESNIL DU BUISSON, F. and SIGNORET, J.P. (1961). *Annls Zootech* **11**, 53
DUNCAN, D.L. and LODGE, G.E. (1960). *Communs Bur. Anim. Nutr., Tech. No. 21.* Aberdeen, Rowett Research Institute
DZIUK, P.J. (1964). In *Conference on estrous cycle control in domestic animals, Nebraska Misc. Publ.* **1005**, 50
DZUIK, P.J. and GEHLBACH, G.D. (1966). *J. Anim. Sci.* **25**, 410
ETIENNE, M. and LEGAULT, C. (1974). *Journées de la recherche porcine en France*, pp. 52–62. Paris, L'Institut Technique du Porc
FOOTE, W.C. and MATTHEWS, D.H. (1969). *J. Anim. Sci.* **29**, 189 (abstract)
FOOTE, W.C., WALDORF, D.P., CHAPMAN, A.B., SELF, H.L., GRUMMER, R.H. and CASIDA, L.E. (1956). *J. Anim. Sci.* **15**, 959
FOWLER, S.H. and ROBERTSON, E.L. (1954). *J. Anim. Sci.* **13**, 949
FRIEND, D.W. (1973). *J. Anim. Sci.* **37**, 701
FRISCH, R.E. and REVELLE, R. (1970). *Science*, N.Y. **169**, 397
GOLDMAN, B.D., GRAZIA, Y.R., KAMBERI, I.A. and PORTER, J.L. (1971). *Endocrinology* **88**, 771
GOODE, L., WARNICK, A.C. and WALLACE, H.D. (1961). *J. Anim. Sci.* **20**, 971 (abstract)
GOODE, L., WARNICK, A.C. and WALLACE, H.D. (1965). *J. Anim. Sci.* **24**, 959
GOSSETT, J.W. and SORENSEN, A.M. Jr. (1959). *J. Anim. Sci.* **18**, 40

HACKER, R.R., KING, G.J. and BEARSS, W.H. (1974). *J. Anim. Sci.* **39**, 155 (abstract)

HAFEZ, E.S.E. (1960). *J. agric. Sci., Cambs.* **54**, 170

HAGINO, N., RAMALEY, J.A. and GORSKI, R.A. (1966). *Endocrinology* **79**, 451

HAINES, C.E., WARNICK, A.C. and WALLACE, H.D. (1959). *J. Anim. Sci.* **18**, 347

HOLLANDBECK, R., BAKER, B.Jr, NORTON, H.W. and NALBANDOV, A.V. (1956). *J. Anim. Sci.* **15**, 418

HOOGSTRA, M.J. and DE JONGH, S.E. (1955). *Acta physiol. phamac. néel.* **4**, 145

HUGHES, E.H. (1934). *Agric. Res., Lond.* **49**, 943

HUGHES, P.E. (1976). PhD Thesis, University of Nottingham

HUGHES, P.E. and COLE, D.J.A. (1975). *Anim. Prod.* **21**, 183

HUGHES, P.E. and COLE, D.J.A. (1976). *Anim. Prod.* **23**, 89

HUGHES, P.E. and COLE, D.J.A. (1978). *Anim. Prod.* **27**, 11

JOHNSON, H.H.K., MOUSTGAARD, J. and HØJGAARD-OLSEN, N. (1952). *Anim. Breed. Abstr.* **23**, 784

KANNWISCHER, R., WAGNER, J. and CRITCHLOW, V. (1967). *Anat. Rec.* **157**, 268

KATO, J., ATSUMI, Y. and INUBA, M. (1974). *Endocrinology* **94**, 309

KELCH, R.P. (1974). In *The endocrine milieu of pregnancy, puerperium and childhood.* 3rd Ross Conference on Obstetric Research

KELCH, R.P., GRUMBACH, M.M. and KAPLAN, S.L. (1972). In *Gonadotrophins* (Ed. by B.B. Saxena). New York, John Wiley and Sons

KING, D.F. (1961). *Poult. Sci.* **40**, 479

KLOTCHKOV, D.V., KLOTCHKOVA, A. Ya., KIM, A.A. and BELYAEV, D.K. (1971). *Proc. 10th int. Congr. Anim. Prod.* (Paris), pp. 1–8

LAZAUSKAS, V.M. (1965). *Anim. Breed. Abstr.* **33**, 3559

LEVINE, S. (1962). In *Roots of Behaviour* (Ed. by Bliss). New York, Harper

LODGE, G.A. and MacPHERSON, R.M. (1961). *Anim. Prod.* **3**, 19

MAJERCIAK, P., SMIDT, D., SCHAHIDI, R. and HARMS, E. (1970). *Z. Tierzuch. Zuchtungsbiol* **86**, 215

MANDL, A.M. and ZUCKERMAN, S. (1952). *J. Endocrinol.* **8**, 357

MARTINAT, F., LEGAULT, C., DU MESNIL DU BUISSON, F., OLLIVIER, L. and SIGNORET, J.P. (1970). *Journées de la recherche porcine en France, 1970*, pp. 47–54. Paris, L'Institut Technique du Porc

MEYER, H.H. and BRADFORD, G.E. (1974). *J. Anim. Sci.* **38**, 271

MIRSKAIA, L. and CREWE, F.A.E. (1930). *Quart. J. exp. Physiol.* **20**, 299

MONTEIRO, L.S. and FALCONER, D.S. (1966). *Anim. Prod.* **8**, 179

MOON, H.D. (1937). *Proc. Soc. exp. Biol., N.Y.* **37**, 34

MORTON, J.R.C., DENENBERG, V.H. and ZARROW, M.X. (1963). *Endocrinology* **72**, 439

MOTTA, M., FRASCHINI, F., GIULIANI, G. and MARTINI, L. (1968). *Endocrinology* **83**, 1101

O'BANNON, R.H. (1964). *Diss. Abstr.* **25**, 5

O'BANNON, R.H., WALLACE, H.D., WARNICK, A.C. and COOMBS, G.E. (1966). *J. Anim. Sci.* **25**, 706

PAREDIS, F. (1961). *Anim. Br. Abstr.* **30**, 2697

PARLOW, A.F., ANDERSON, L.L. and MELAMPY, R.M. (1964). *Endocrinology* **75**, 365

PAY, M.G. and DAVIES, T.E. (1973). *Anim. Prod.* **17**, 85

PEDERSON, T. (1969). *Acta Endocr. Copenh.* **62**, 117

PHILLIPS, R.W. and ZELLER, J.H. (1943). *Anat. Rec.* **85**, 387

PLUMLEE, M.D., THRASHER, D.M., BEESON, W.W. ANDREWS, F.N. and PARKERS, H.E. (1956). *J. Anim. Sci.* **15**, 352

PRESL, J., HERZMANN, J. and HORSKY, J. (1969). *J. Endocr.* **45**, 611

PRICE, D. and ORTIZ, E. (1944). *Endocrinology* **34**, 215

RAMIREZ, V.D. (1973). In *Handbook of physiology* (Ed. by S.R. Geiger), *Section 7: Endocrinology*, Vol. 2, Part 1

RAMIREZ, V.D. and McCANN, S.M. (1963). *Endocrinology* **72**, 452

RAMIREZ, V.D. and SAWYER, C.H. (1965). *Endocrinology* **76**, 412

REDDY, V.B., LASLEY, J.F. and MAYER, D.T. (1958). *Res. Bull. Mo. agric. Exp. Stn*, No. 666

REUTZEL, L.F. and SUMPTION, L.J. (1968). *J. Anim. Sci.* **27**, 27

ROBERTSON, G.L., CASIDA, L.E., GRUMMER, R.H. and CHAPMAN, A.B. (1951a). *J. Anim. Sci.* **10**, 841

ROBERTSON, G.L., GRUMMER, R.H., CASIDA, L.E. and CHAPMAN, A.B. (1951b). *J. Anim. Sci.* **10**, 647

RUIZ, M.E., SPEER, V.C., HAYS, V.W. and SWITZER, W.P. (1968). *J. Anim. Sci.* **27**, 1602

SCHMIDT, K. and BRETSCHNEIDER, W. (1954). *Anim. Breed. Abstr.* **22**, 1046

SCHOELLER, V., DOHRN, M. and HOHLWEG, W. (1936). *Klin. Wschr.* **15**, 1907

SCHWARTZ, F.L., ROBISON, O.W. and ULBERG, L.C. (1971). *J. Anim. Sci.* **32**, 391 (abstract)

SELF, H.L., GRUMMER, R.H. and CASIDA, L.E. (1955). *J. Anim. Sci.* **14**, 573

SHORT, R.E. and BELLOWS, R.A. (1971). *J. Anim. Sci.* **32**, 127

SHORT, R.E., ZIMMERMAN, D.R. and SUMPTION, L.J. (1963). *J. Anim. Sci.* **22**, 868 (abstract)

SIGNORET, J.P. (1970). *J. Reprod. Fert., Suppl.* **11**, 105

SMITH, E.R. and DAVIDSON, J.M. (1968). *Endocrinology* **82**, 100

SORENSEN, A.M. Jr., THOMAS, W.B. and GOSSETT, J.W. (1961). *J. Anim. Sci.* **20**, 347

SQUIERS, C.D., DICKERSON, G.E. and MAYER, D.T. (1950). *J. Anim. Sci.* **9**, 683 (abstract)

SURMUHIN, A.F. and CEREMNYH, V.D. (1970). *Anim. Breed. Abstr.* **39**, 3687

SVIBEN, M., SALEHAR, A., HERAK, M., CERNE, F. and VINOURSKI, Z. (1969). *Anim. Breed. Abstr.* **37**, 3897

WARNICK, A.C., WIGGINS, E.L., CASIDA, L.E., GRUMMER, R.H. and CHAPMAN, A.B. (1951). *J. Anim. Sci.* **10**, 479

WHETTEMAN, R.P., OMTVEDT, I.T. and WILLIAMS, T.W. (1973). *Res. Rep. agric. Exp. Stn, Okl. St. Univ.* **(1973) MP – 90**, 181

WITZ, W.M. and BEESON, W.M. (1951). *J. Anim. Sci.* **10**, 957

WURTMAN, R.J. (1967). In *Neuroendocrinology* (Ed. by W. F. Martini and L. Ganong). New York, Academic Press

WURTMAN, R.J., AXELROD, J. and CHU, E.W. (1963). *Science, N.Y.* **141**, 277

YAGIMURA, T., KIGAMA, T. and KOBAYASHI, K. (1969). *J. Endocr.* **45**, 473

ZARROW, M.X., CLARK, J.M. and DENENBERG, V.H. (1969). *Neuroendocrinology* **4**, 270

ZIMMERMAN, D.R., CARLSON, R. and NIPPERT, L. (1969). *J. Anim. Sci.* **29**, 203 (abstract)

ZIMMERMAN, D.R., SPIES, H.G., RIGOR, E.M., SELF, H.L. and CASIDA, L.E. (1960). *J. Anim. Sci.* **19**, 687

Chapter 4

The oestrous cycle

The pig does not have a specific breeding season as, for example, does the sheep. Instead, it is fertile throughout the year, regular oestrous cycles occurring approximately every 21 days (range 18–23 days). These cycles commence immediately following the attainment of puberty and continue throughout the female's life, interrupted only by pregnancy and lactation. They are characterized by periods of reproductive quiescence lasting 18–20 days followed by short periods of sexual receptivity, having a mean duration of 53 hours (range 12–72 hours). These interludes of sexual receptivity are known as the oestrus, or heat periods. At this time the vulva becomes swollen and reddens, and this is often accompanied by a discharge of mucus from the vulva. The oestrous female will go in search of a boar and will stand close to him in readiness for mating. She will exhibit the 'standing heat reflex' (i.e. will stand immobile in the proximity of a boar, or if pressure is applied to her back) and, in the case of breeds with prick ears (e.g. Large White), will cock her ears. The mature ova will be released from the ovary during the second half of the oestrous period, normally within the range 38–42 hours after the onset of oestrus. The total time taken for ovulation to occur is approximately 3.8 hours (Signoret, 1972), with the range in ovulation rate of 10–25 ova.

4.1 Ovarian and uterine changes

The ovary of the female pig begins life with a large pool of germ cells, each having the potential to develop into an ovulatory follicle. The size of this pool is, in fact, far in excess of the requirements of the animal. From this pool germ cells are continually being initiated (the mechanism whereby selection is made is unknown) into a growth sequence which, once begun, continues until the follicles ovulate or become atretic (Richards, Rao and Ireland, 1978). Evidence from the rat and mouse suggests that, in these species at least, the time taken for a small follicle which has entered the pool of 'committed growing follicles' to develop to the stage of an antral, pre-ovulatory follicle is four oestrous cycles (Richards, Rao and Ireland, 1978). In fact, most of these follicles will only develop as far as the large pre-antral follicle stage and will then undergo atresia. The few follicles which actually enter the final stages of growth appear to be 'selected' from the pool of large, pre-antral follicles at an early stage in the cycle preceding ovulation. At the beginning of that cycle they will still be large, pre-antral follicles with a diameter no greater than 4 mm. Follicular size will then increase slowly over the first 16 days of the cycle to about 6 mm, final maturation to antral, pre-ovulatory follicles with a diameter of 10–12 mm occurring prior to ovulation. These stages of follicular growth are shown diagrammatically in *Figure 4.1*.

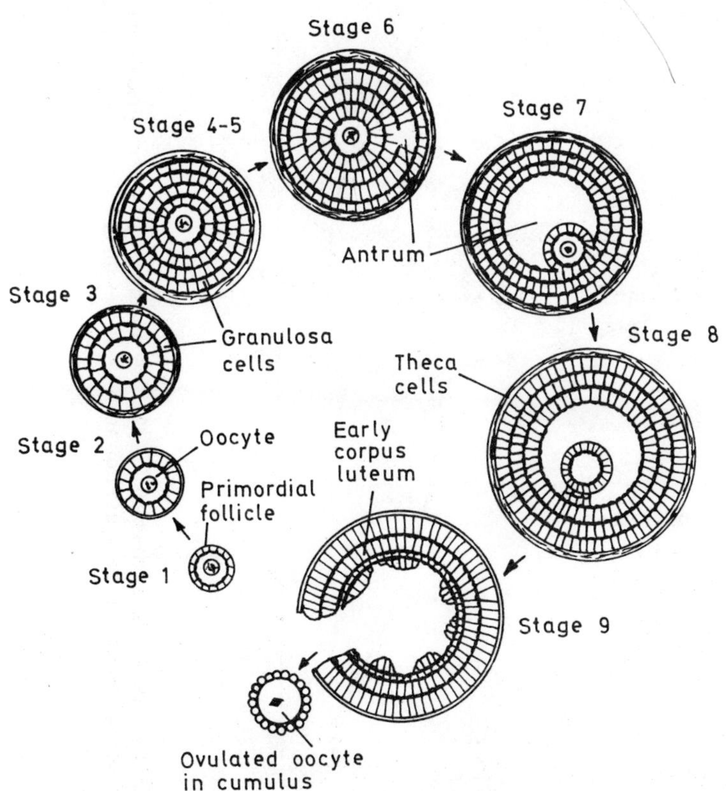

Figure 4.1 A diagrammatic representation of follicular growth (from Baker, 1973)

Immediately following ovulation the walls of each ruptured follicle collapse, usually around a central blood clot, and the diameter reduces to 4–6 mm. The remaining granulosa cells of the ruptured follicle then hypertrophy to become the lutein cells of the forming

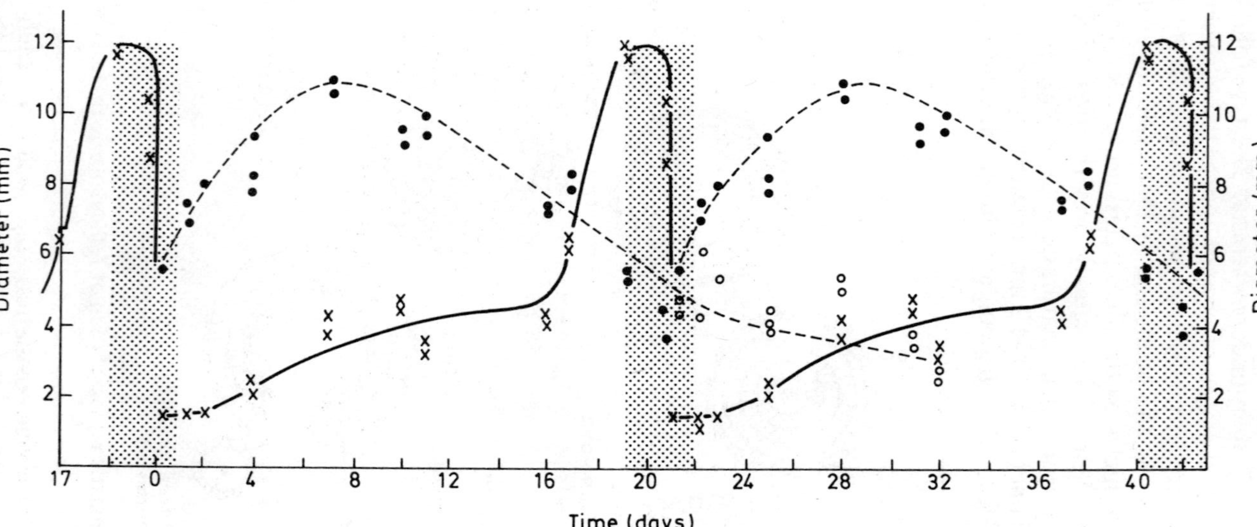

Figure 4.2 Changes occurring in the ovary of the sow during the oestrous cycle (from Hammond, Mason and Robinson, 1971). The stippled areas mark the period of heat. ——— follicles; – – – – – corpora lutea

corpus luteum. The theca interna cells multiply and migrate into the corpus luteum where they are invaded by a network of capillaries (Anderson and Melampy, 1967). Thus formed, the corpus luteum rapidly increases in size, such that it has a diameter of 8–9 mm by day 7 of the cycle. Little morphological change then occurs until day 14 or 15 of the cycle. At this time, assuming no fertilization has occurred, there is a rapid decline in corpus luteum diameter to about 6 mm over a 2–3 day period. This is associated with the complete breakdown of the lutein cells and the collapse of the accompanying capillaries. Visually this change is seen as a regression from the pink active corpus luteum of the early stages of the cycle to the white scar tissue of the regressing inactive corpus luteum. Subsequently, these inactive corpora lutea further degenerate to small masses of scar tissue, known as corpora albicans. These changes in corpus luteum development, together with follicular growth, are represented in *Figure 4.2*.

Concurrent with these ovarian developments are morphological changes in the uterus of the female. During the follicular phase of the cycle (i.e. from day 16 until ovulation) the uterine endometrium forms a relatively thin layer, and the glands of the uterus tend to be simple and straight, with few branchings. However, once ovulation has occurred and the luteal phase of the cycle initiated, the uterine endometrium displays a conspicuous increase in thickness. Furthermore, the uterine glands grow rapidly in both diameter and length, becoming extremely branched and convoluted (Nalbandov, 1964). These postovulatory changes occur in order to prepare the uterus for ova that may be fertilized during the oestrous period (a subject dealt with in greater detail in Chapters 5 and 6). However, if no fertilization occurs the thick endometrium and complex uterine glands become unnecessary. Once the corpora lutea become inactive (at the end of the luteal phase of the cycle), the excess uterine endometrium is sloughed off and excreted.

4.2 Endocrine status

The morphological changes described above are controlled by circulating hormones, these originating from the anterior pituitary gland, ovary and uterus. In recent years the development of radioimmunoassays has enabled measurement of the concentrations of these hormones in the blood of the female and from this basic knowledge a pattern of hormone secretion throughout the oestrous cycle has been established. Furthermore this information has been translated into a hypothesis of the mechanisms involved in the control of the oestrous cycle, an outline of which is given below. It should however be stressed that much of the information on which the hypothesis is based has been obtained from laboratory animals, and, as such, must be subject to critical examination.

THE OVARIAN HORMONES

The ovary secretes the two steroid hormones oestrogen and progesterone produced by the growing follicles and the corpora lutea respectively – these hormones are also secreted in small quantities by the adrenal glands. The pattern of secretion of the ovarian hormones through the oestrous cycle is shown in *Figure 4.3*.

This demonstrates that circulating oestrogen levels are low for most of the oestrous cycle, but begin to rise from about day 17 onwards to a maximum level on day 19 or 20. This rise occurs at a time when follicular growth and maturation is at a maximum and culminates in the pre-oestrus oestrogen peak. Thus, it would appear that the secretion rate of oestrogen is largely dependent on the rate of growth and degree of maturity of the ovarian follicles. Furthermore it should be added that the high circulating levels of oestrogen are primarily responsible for the behavioural patterns and vulval changes seen just prior to, and during, the oestrous period.

Following ovulation the major sources of oestrogen, namely the large antral follicles, are converted into the corpora lutea of the luteal phase of the cycle. As these develop they begin to secrete progesterone in increasing quantities. *Figure 4.3*(*c*) shows that the level of release of progesterone may not, in fact, reach a maximum until as late as day 16 of the cycle. Moreover, it should be emphasized that circulating progesterone levels are extremely variable in both pattern and level. Thus, for example, the peak level of plasma progesterone may vary from 7.5 ng/ml to 56.1 ng/ml and occur anywhere between days 8 and 16 of the cycle without affecting normal cyclicity (Hughes, 1976).

THE UTERINE HORMONES

The uterus does in fact only produce one known hormone that influences the oestrous cycle. This hormone, prostaglandin, is released from the uterus throughout the cycle, but shows one distinct peak just prior to the end of the luteal phase (*Figure 4.3* (*b*)). The prostaglandin is carried in the uterine vein and is transferred to the ovarian artery via a counter-current exchange system. This localized transfer of the hormone is a necessary development if the prostaglandin is to be active on reaching the ovary, as it would be rapidly broken down in the lungs if it had to travel to the ovaries via the general circulation.

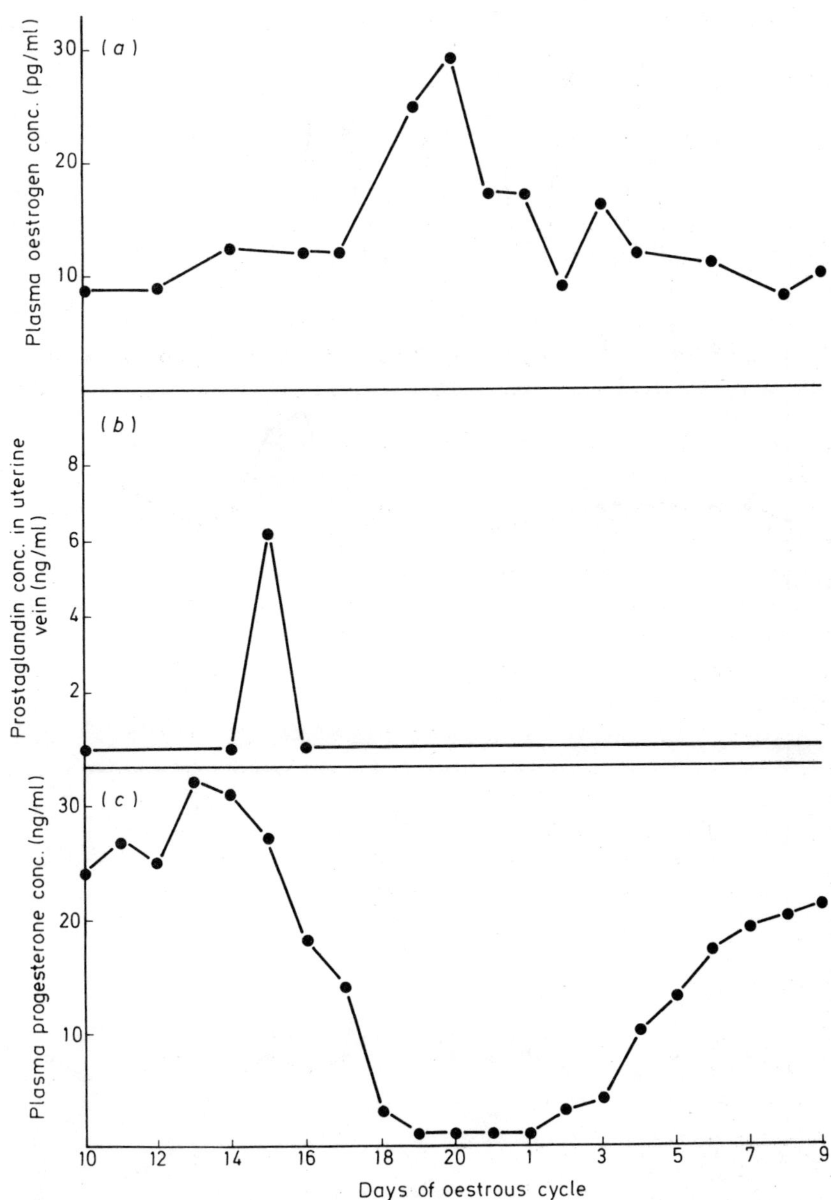

Figure 4.3 The pattern of oestrogen and progesterone secretion during the oestrous cycle of the pig (from Stabenfeldt *et al.*, 1969; Edqvist and Lamm, 1971; Guthrie, Hendricks and Handlin, 1972; Shearer *et al.*, 1972)

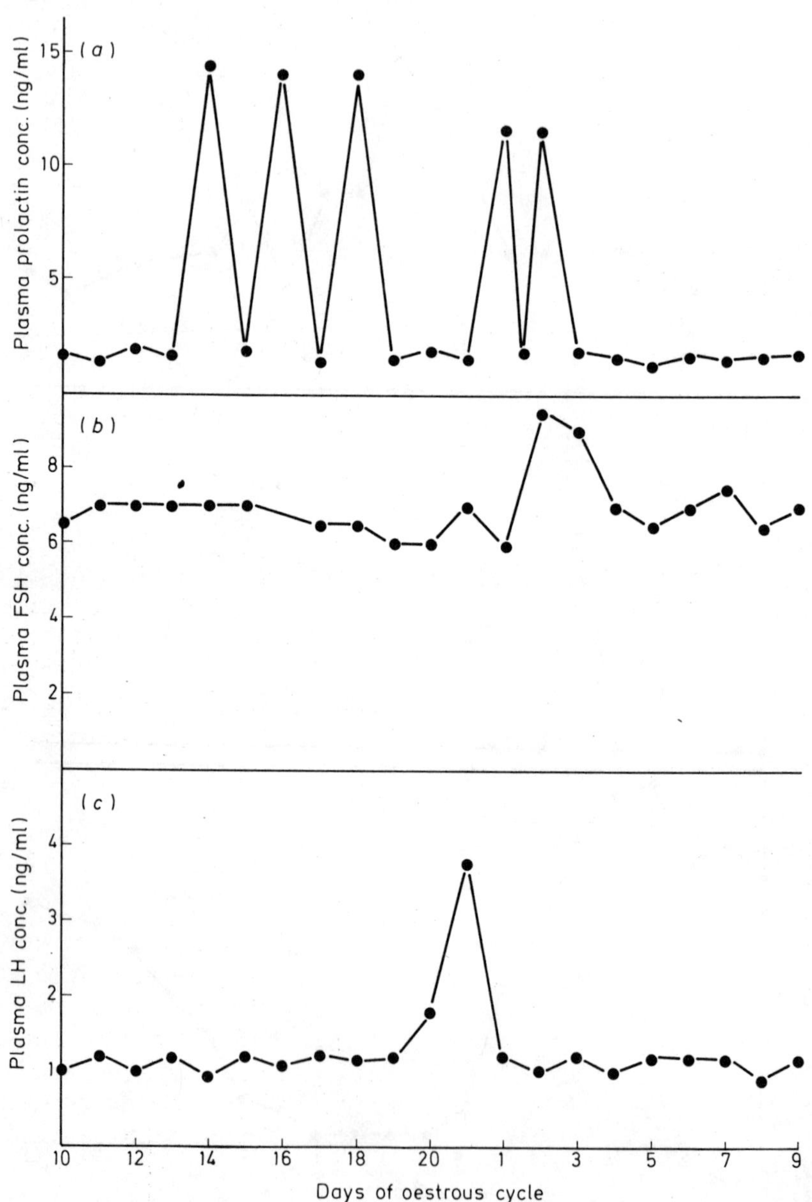

Figure 4.4 The pattern of FSH, LH and prolactin secretion during the oestrous cycle of the pig (from Rayford, Brinkley and Young, 1971; Hendricks, Guthrie and Handlin, 1972; Brinkley, Wilfinger and Young, 1973; Wilfinger, Brinkley and Young, 1973a, b)

THE PITUITARY HORMONES

The three hormones secreted by the anterior pituitary gland which are of prime importance in the control of the oestrous cycle are LH, FSH and prolactin. The circulating levels of these hormones during the cycle are given in *Figure 4.4*.

It is clear that LH secretion is at a minimum for the majority of the cycle, with only one significant peak of release being achieved. This peak follows the pre-oestrus oestrogen peak described above, and occurs just prior to ovulation.

On the other hand FSH displays a slightly more erratic secretion pattern. The circulating levels of this hormone are fairly low for most of the oestrous cycle, but they do show two distinct peaks about the time of ovulation. The first and smaller peak occurs at the same time as the pre-ovulatory LH peak. The second FSH peak is considerably larger and occurs on day 2–3 of the cycle.

The secretion pattern for prolactin is less well known. However, it has been reported that a peak of prolactin release occurs after the pre-ovulatory LH peak, followed by a second peak on day 2 of the cycle (Brinkley, Wilfinger and Young, 1973). This is shown in *Figure 4.4(a)*.

It is clear from *Figure 4.4(a)* that other peaks of prolactin are also present during the oestrous cycle. This is also true of the hormones FSH and LH. Wilfinger, Brinkley and Young (1973a, b) have reported that an average of 8.5 secondary LH peaks (with an average maximum plasma concentration of 2.34 ng/ml) occur during the luteal phase of the cycle, and at least one secondary FSH peak (this being associated with a secondary LH peak).

4.3 Control mechanisms in the oestrous cycle

FOLLICULAR DEVELOPMENT

It has already been stated that the few follicles which enter the final stages of growth are 'selected' from a pool of large, pre-antral follicles at an early stage in the cycle preceding their ovulation. Once ovulation has occurred this selection procedure is initiated. The actual selection mechanism is unknown, although it does seem likely that the peak of FSH release occurring on days 2–3 of the cycle is involved. Indeed, this may be the sole 'selection' stimulus, or there could possibly be a synergistic action of FSH and prolactin causing 'selection'.

Once selection has occurred the chosen follicles must grow and develop until they are mature and ready to be ovulated at the next oestrus. The gonadotrophin hormones, FSH and LH, are primarily responsible for the stimulation of this follicular growth. However, during the luteal phase of the cycle, the high circulating levels of progesterone prevent the secretion of large quantities of the gonadotrophins. Since this negative feedback effect of progesterone continues until the end of the luteal phase (i.e. about day 16 of the cycle), follicular growth is minimal up to this stage of the oestrous cycle. However, between days 16 and 21 of the cycle follicular growth increases markedly. It is therefore surprising that little change is seen in the rate of secretion of FSH and LH at this time. It must be speculated that the ovaries become more sensitive to gonadotrophic stimulation at this time, thus allowing follicular growth to occur in response to relatively low levels of circulating gonadotrophin. This change in ovarian sensitivity may well be due to the change in steroid balance, from progesterone domination to oestrogen domination, at the beginning of the follicular phase of the cycle.

During the follicular phase gonadotrophic stimulation of the follicles causes increased secretion of oestrogen. This rise in the circulating level of oestrogen is slow at first, but builds up to culminate in a pre-oestrus surge on the day prior to ovulation. This probably occurs in response to a change in the feedback effects of oestrogen, from negative to positive, once a pre-optic threshold level for oestrogen has been reached (*see* Chapter 2). Thus, follicular growth is minimal during the luteal phase, increases during the follicular phase, and shows a rapid 'maturation spurt' just prior to ovulation. This final burst of growth results from the positive feedback of oestrogen causing high secretion rates of FSH and LH. Finally, this increase in gonadotrophin levels results in a pre-ovulatory surge of LH, and probably FSH, responsible for the actual release of the ova from the follicles (i.e. ovulation).

CORPUS LUTEUM DEVELOPMENT

Once ovulation has occurred the ruptured follicles must luteinize to form corpora lutea. This luteinization is thought to occur in response to high circulating levels of LH. Once formed, the corpora lutea begin actively secreting progesterone, continuing until luteolysis occurs at approximately day 16 of the cycle. Indeed, progesterone secretion would continue for about 113 days if a pregnancy was established. The factor which causes luteolysis in the non-pregnant female is now known to be prostaglandin. This uterine hormone shows one distinct peak of release on day 15–16 of the cycle, coincident with the initiation of the decline in progesterone secretion. The prostaglandin acts at the ovarian level to cause luteolysis of the corpora lutea, and therefore a rapid decline in circulating progesterone levels. In the pregnant female this does not occur, although the release of prostaglandin is still apparent. In this case oestrogen secretion from the embryos causes the prostaglandin to be released into the lumen of the uterus and not into the uterine vein, and hence the prostaglandin does not reach the ovary to produce its luteolytic effects.

The above theory of control mechanisms in the oestrous cycle of the female pig is, as has already been emphasized, one possible explanation of the events that occur. Many of the events described are now accepted as fact while others, such as the method of selecting follicles for further development, are yet to be confirmed. Indeed, it should also be added that the purposes of many hormonal releases are yet to be discovered. For example, the reasons for the secretion pattern of prolactin, given in *Figure 4.4(a)*, have not been identified. The two surges of this hormone that occur at about the time of ovulation may be associated with either the luteinization of the ruptured follicles, maintainance of the corpora lutea (although this is reported to be unlikely (Anderson, 1966)), selection of follicles for the next ovulation or, possibly, some other undefined action. Other surges of prolactin occurring during the oestrous cycle have yet to be attributed a function. Indeed, the same may also be said for the peaks of LH and FSH occurring during the luteal phase of the cycle. These may possibly have a luteotrophic function or, alternatively, may simply be an overflow from a gonadotrophin pool at the pituitary level.

Our knowledge of the control of the female oestrous cycle is therefore incomplete. Most of the changes that occur may now be satisfactorily explained, but others for the present remain a mystery. However, the foregoing description provides a basis on which to explain the changes that occur in response to the attempted manipulation of the cycle of the female pig.

4.4 Artificial control of the oestrous cycle

The principle reason for artificially controlling the oestrous cycle of the pig is to synchronize the onset of oestrus in groups of females. This has the obvious advantage of allowing breeding pigs to be treated as groups rather than as individuals. In this situation groups, or batches, of sows would be mated at the same time, farrow together, and subsequently be weaned as a group. Thus, not only would the sows be considered as a batch but also their offspring would be of similar size, hence allowing easy mixing and rearing. The other two principle advantages of synchronizing oestrus are that the employment of artifical insemination is facilitated (indeed it may be necessitated since groups of females will require mating on the same day) and superovulation may be employed.

The two methods whereby oestrus may be synchronized are by delaying the onset of heat or by prematurely inducing it. Oestrus inhibition or delay is usually accomplished by extending the luteal phase of the cycle via the administration of progesterone or synthetic progestagens. On the other hand, a variety of methods have been used in an attempt to induce oestrus during the cycle. These include treatment with PMS/HCG, FSH, GnRH, and prostaglandins.

The efficiency of these treatments aimed at either delaying or inducing oestrus are discussed below.

DELAYING THE ONSET OF OESTRUS

The reasoning behind delaying oestrus is that all the treated females attain what is effectively the same physiological state. Thus, when the delaying influence is withdrawn the females should all take approximately the same time to return to oestrus. The easiest way in which oestrus and ovulation may be inhibited is by the suppression of gonadotrophin release. In theory this may be achieved by either maintaining the corpora lutea beyond the normal point of regression, by providing the equivalent of corpora lutea via the administration of progesterone or synthetic progestagens, or by administering compounds which simply inhibit gonadotrophin release.

Maintenance of the corpora lutea

Little convincing evidence is available concerning the maintenance of corpora lutea beyond the normal point of regression. Administration of progesterone or progestagens does not extend the life of the corpora lutea, although it does inhibit gonadotrophin release (*see below*). Indeed, it is clear that the effects of prostaglandin on the corpora lutea are dependent on the age of the corpora lutea and not on progesterone levels (Garbers and First, 1969). Thus, irrespective of progesterone concentration in the circulation, the luteolytic effects of prostaglandin will occur at approximately day 16 of the cycle. The exception to this rule appears to be the oestrogen-treated female. Indeed, many reports from Eastern Europe suggest that the administration of high levels of oestrogen can suppress oestrus activity in the sow for several months (Bajez, 1952; Benesch, 1953; Kment and Halama, 1953; Sporri and Candinas, 1951; Stift, 1953; Zehetner, 1953). Furthermore, Kidder, Casida and Grummer (1955) observed that relatively low levels of diethylstilboestrol (3 mg) could also prolong the oestrous cycle by as much as 6 days when administered on day 11 of the cycle. In this experiment administration of oestrogen on day 6 of the cycle had no effect, whereas a variable response was obtained when treatment commenced on day 16. However, it should be added that these workers did observe that some oestrogen-treated gilts had abnormally large luteinized follicles following treatment.

The delay in the onset of oestrus when gilts are treated with oestrogen on day 11 of the cycle appears to be due to a prolongation of the life of the corpora lutea rather than just the suppression of gonadotrophin release (Gardner, First and Casida, 1963). Indeed, most of the work carried out using oestrogen treatment of cycling females has been based on the concept that oestrogen prolongs the life of the corpora lutea by stimulating the release of endogenous luteotrophic hormones (usually thought to be prolactin, since this has been identified as being luteotrophic in the rat). However, more recent evidence suggests that corpus luteum function in the pig is independent of luteotrophic support once luteinization has been initiated by the pre-ovulatory surge of LH (Du Mesnil du Buisson, 1966). If this is proved to be the case then the mechanism whereby exogenous oestrogen prolongs the life of the corpora lutea is difficult to identify. It may be that the corpora lutea are more sensitive to luteotrophic support (either from prolactin or LH) towards the end of the luteal phase of the cycle. Alternatively, it is possible that the exogenously administered oestrogen produces its effects at the uterine level, emulating the role of embryonic oestrogen in redirecting prostaglandin flow from the uterine vein to the lumen of the uterus. However, irrespective of its mechanism of action, the use of exogenous oestrogen as a means of controlling the oestrous cycle of the pig is limited by two factors – first, it is only effective during a short period of the cycle, and secondly it may result in the development of cystic follicles.

Suppression of gonadotrophin release

Most attempts to suppress gonadotrophin release, excepting those involving extension of the life span of the corpora lutea, have involved the use of progesterone or synthetic progestagens. The administration of these compounds provides the female with an artificial luteal phase, hence preventing any major release of gonadotrophins. This is due to the negative feedback influence of progesterone on GnRH release from the hypothalamus.

As early as 1951, Ulberg, Grummer and Casida investigated the effects of administering progesterone to cycling gilts. This work showed that oestrus and ovulation could be suppressed if a minimum level of 25 mg/day progesterone was given to gilts early enough in the cycle. However, they also report that gilts treated with progesterone tended to have a high incidence of cystic follicles. Subsequent work has confirmed the observation that oestrus is suppressed when a level of 25 mg/day, or above, of progesterone is given to the gilt (Baker *et al.*, 1954; Dzuik, 1960; Gerrits *et al.*, 1963), although Dzuik (1960) reported that ovulation was not blocked when a level of 1000 mg/day was administered (*Table 4.1*). Most progesterone-treated females

Table 4.1. The effects of exogenous progesterone administration on reproductive performance in gilts and sows

Author	*Animals used*	*Time progesterone applied*	*Progesterone dose*	*Effects*	*Cystic follicles present* (+) *or absent* (–)
Ulberg *et al.* (1951)	Cycling gilts	–	12.5 mg/day	Oestrus and ovulation inhibited when 25 mg/day and above used	+
Baker *et al.* (1954)	Cycling gilts	Day 10 or 15 of the cycle until day 28	25 mg/day 100 mg/day	Oestrus and ovulation inhibited by both treatments. Subsequent fertility low	+
Dzuik (1960)	Cycling gilts/ weaned sows	From weaning or any time in the cycle, for 14–21 days	1000 mg/day	Oestrus inhibited in only 62.5 % of treated animals. Ovulation not inhibited. Subsequent fertility normal	–
Gerrits *et al.* (1963)	Cycling gilts	From days 4 to 10 of cycle	100 mg/day 300 mg/3 days	Oestrus and ovulation inhibited by both treatments. Subsequent fertility normal	–

have been reported to return to oestrus 4–8 days following the withdrawal of treatment (Webel, 1978). The fertility at this oestrus is variable, some reports indicating that normal ovulation and fertilization occur whereas others indicate a sharp decline in subsequent fertility (e.g. Baker *et al.* (1954) reported that the average number of fertilized ova present in progesterone-treated gilts following mating was 5.5 less than for control animals). Also several reports indicate that the incidence of cystic follicles increases markedly in progesterone-treated females (Ulberg *et al.*, 1951; Baker *et al*; 1954).

Another approach to the suppression of gonadotrophin release has been the use of synthetic progestagens. The compounds used include methylacetoxyprogesterone (MAP), chlormadinone acetate (CAP), megestrol acetate (AMP) and more recently allyloestrenol (A-35957). These compounds have been administered at levels varying from 3.25 to 540 mg/day for periods from 7 to 27 days (Dzuik, 1960; Nellor, 1960; Nellor *et al.*, 1961; Wagner and Seerley, 1961; First *et al.*, 1963; Pond *et al.*, 1965; Veenhuizen *et al.*, 1965; Webel, 1978). Furthermore, they have been given to both weaned sows and to gilts at various stages of the oestrous cycle. The results of these studies indicate that the administration of progestagens to gilts or sows results in the inhibition of oestrus and ovulation, with a return to oestrus 4–6 days following withdrawal of treatment. The optimum treatment appears to be a dose of 50–100 mg/day given over a 15 day period. However, with the possible exception of A-35957, subsequent fertility in progestagen-treated females has been impaired by the high incidence of cystic follicles occasioned by the treatment (*Table 4.2*). Furthermore, the administration of PMS and other ovulation-stimulating substances appears to have little beneficial effect on the fertility of the treated females (Webel, 1978).

Thus, although the use of progestagens, and indeed progesterone, does facilitate control of the oestrus cycle of the pig, it appears likely that their use will be severely limited by the adverse effects on subsequent fertility. In particular, the high incidence of cystic follicles and the reduced rate of fertilization (Baker *et al.*, 1954) render the employment of these compounds unwise. The low fertilization rates encountered are likely to be due to progesterone treatment disturbing the steroid balance at and around the time of mating, since evidence suggests that the correct oestrogen/progesterone balance is essential at this time to ensure normal egg transport and the maintenance of an optimum uterine environment (Day and Polge, 1968). The occurrence of cystic follicles following progesterone or progestagen treatment appears to be due to failure to completely inhibit the release of gonadotrophins during treatment. It seems likely that administered progesterone/progestagen only inhibits the release of large quantities of gonadotrophins (i.e. surge-type release), while

Table 4.2. The effects of various synthetic progestagens on reproductive performance in gilts and sows

Author	*Animals used*	*Time progestagen applied*	*Progestagen type*	*Progestagen dose*	*Effects*	*Cystic follicles present* (+) *or absent* (–)
Dzuik (1960)	Cycling gilts/ weaned sows	From weaning or at any time in the cycle, for 14–21 days	MAP	20–300 mg/day	All dose levels inhibited oestrus. Ovulation inhibited above 50 mg/day Subsequent fertility normal	–
Nellor (1960)	Cycling gilts	From days 11 to 16 of the cycle	MAP	1.76–3.52 mg/kg body wt/day	All dose levels inhibited oestrus and ovulation. Subsequent fertility normal	–
Nellor *et al.* (1961)	Cycling gilts	For 7–27 days starting at various times during the cycle	MAP	0.88–3.52 mg/kg body wt / day	All dose levels inhibited oestrus and ovulation (optimum was 0.5 mg for 15 days). Subsequent fertility normal	–
Wagner and Seerley (1961)	Cycling gilts	For 18 days starting at various times during the cycle	CAP	3.25–540 mg/ day	Oestrus and ovulation inhibited when 25 mg/day and above used. Subsequent fertility low	+

First *et al.* (1963)	Cycling gilts/ weaned sows	(Various)	MAP	50–400 mg/ day	Oestrus and ovulation inhibited when 100 mg/ day and above used. Subsequent fertility low	+
Pond *et al.* (1965)	Cycling gilts	For 15 days starting at various times during the cycle	AMP	0.66–1.1 mg/ kg body wt / day	All dose levels inhibited oestrus and ovulation. Subsequent fertility low	+
Veenhuizen *et al.* (1965)	Cycling gilts	For 10 days after 25 mg DES*/day for 10 days	CAP	50 mg/day	Oestrus and ovulation inhibited. Subsequent fertility low	+

*DES = diethylstilboestrol.

allowing small episodic releases to continue (as, indeed, they do during the luteal phase of a normal oestrous cycle). Thus, follicular growth does continue, albeit slowly, thoughout the period of oestrus inhibition. The result of this is that follicular development continues beyond the normal ovulatory size and the follicle eventually becomes cystic (Garbers and First, 1969). Hence, any treatment which is aimed at inhibiting oestrus and ovulation in the pig must afford complete suppression of gonadotrophin release during the treatment period if the above problems are to be avoided.

The only compound that has been found which meets these requirements is a non-steroid compound called methallibure (also known as ICI 33828 or Aimax). This compound could be mixed with

Table 4.3. The effects of methallibure (ICI 33828) on reproductive performance in the gilt

Author	*Period of application*	*Dose*	*Inhibition of oestrus and ovulation*	*Subsequent fertility*
Gerrits and Johnson (1965)	From day 1 to 17 of the cycle for 19 days	0.9–2.14 mg/kg body wt / day	Total	Normal
Stratman and First (1965)	From day 3, 11 or 18 of the cycle for 16 days	58 mg/day 116 mg/day 232 mg/day	Total	Normal
Groves (1967)	20 days	100 mg/day	Total	Normal
Baker *et al.* (1970)	20 days	100 mg/day	Total	Normal
Webel *et al.* (1970)	20 days	–	Total	Normal

the animals' feed and thus be provided daily. It successfully inhibited oestrus and ovulation while being administered, and resulted in the appearance of an oestrus which was of normal fertility 4–9 days following withdrawal. *Table 4.3* cites a few of the many experiments which illustrate the potential value of methallibure. Unfortunately, this compound has now been withdrawn from the market by the manufacturers following reports of teratogenic effects occurring in pregnant gilts treated with it. Thus, its availability and use are unlikely to be widespread in the future.

THE PRECOCIOUS INDUCTION OF OESTRUS

In many cases where oestrus and ovulation have been inhibited by treatment with oestrogen, progesterone, progestagens or methallibure, gonadotrophins (in particular PMS) have been subsequently used in an attempt to increase ovulation rate and give a precise timing of ovulation. The value of this treatment in terms of increasing litter size is minimal (Gibson, *et al.*, 1963; Longnecker and Day, 1968; Christensen *et al.*, 1973), although beneficial effects may be obtained when treating sows with a history of low litter size (Schilling and Cerne, 1972). In contrast, the use of HCG or GnRH to induce ovulation is undoubtedly of value when fixed-time insemination is to be used (Dziuk, Polge and Rowson, 1964; Hunter and Polge, 1966). Thus, these various gonadotrophin preparations may have a role in situations where oestrus inhibition is practised.

However, they have also been employed in attempts to induce precocious oestrus in pigs via their administration during either the luteal or follicular phase of the oestrous cycle. The effects of these treatments are undoubtedly influenced by the stage of the oestrus cycle where they are administered (Tanabe *et al.*, 1949; Day *et al.*, 1959; Polge, 1972). Indeed, the administration of PMS to gilts during the follicular phase of the cycle results in an oestrus of normal fertility 3–4 days later (Hunter, 1966), whereas the same treatment given during the luteal phase of the cycle may not result in spontaneous ovulation, and oestrous exhibition is often suppressed (Hunter, 1967). In addition, if insemination is carried out at this induced oestrus, fertility is generally low, with frequent abnormalities of fertilization and egg transport occurring (Casida, 1965; Hunter, 1967). These adverse effects are attributable to the high circulating levels of progesterone apparent in gilts treated with gonadotrophins during the luteal phase of the cycle (Day and Polge, (1968).

It is clear that the use of gonadotrophin preparations to induce oestrus and ovulation in groups of pigs which are randomly cycling is impractical. The only other available method of inducing a premature oestrus in the cycling pig is to cause regression of the corpora lutea by the administration of prostaglandins. This method of controlling the oestrous cycle has been applied with a high degree of success in both cattle and sheep (c.f. Lamming, Hafs and Manns, 1975; Gordon, 1975). Unfortunately, the corpora lutea of the pig appear to be refractory to prostaglandins until about day 11 or 12 of the cycle (Diehl and Day, 1974; Hallford *et al.*, 1975; Guthrie and Polge, 1976a). Premature regression of pig corpora lutea, therefore, can be elicited by administered prostaglandins only on days 12, 13, 14 or 15 of the cycle. Clearly, this short period of prostaglandin-sensitivity during the cycle does not lend itself to a practical system for oestrous cycle control in the pig. Indeed, the only role for prostaglandins in the artificial control of pig oestrous cycles appears to be their possible use to synchronize the regression of oestrogen-maintained corpora lutea or PMS/HCG-induced accessory corpora lutea (Guthrie, 1975; Guthrie and Polge, 1976b).

4.5 Conclusions

The potential advantages of controlling the oestrous cycle of the pig are fairly clear. Treatment of breeding pigs as groups, facilitation of artificial insemination and superovulation all represent important aids to management and potential stimulants to increased reproductive and economic efficiency. Therefore, the foregoing discussion of the methods available to control the oestrous cycle should be of considerable importance. Unfortunately, the conclusions to be drawn from that discussion are not so helpful.

The cycle may be controlled, and the appearance of oestrus in groups of females synchronized, by either inhibiting oestrus in the females and then removing the suppressant or by stimulating the onset of oestrus in randomly cycling females. However, while oestrus inhibition (using either oestrogens, progesterones or progestagens) may be effectively carried out, subsequent fertility is usually low and the incidence of cystic follicles is significantly increased. Similarly, induction of oestrus during the cycle has resulted in variability in both the appearance of oestrus and subsequent fertility. Thus, while artificial manipulation of the oestrous cycle has been achieved, the results indicate that none of the systems at present available may be satisfactorily employed in commercial practice. What alternative methods of oestrus synchronization are available therefore to the producer?

In Chapter 3, when discussing systems of puberty stimulation in the gilt, it was emphasized that synchronization of the pubertal oestrus was an important criterion in determining the efficiency of any given stimulus. Some of the stimuli employed to induce precocious puberty did also result in a high degree of oestrus synchronization. Prominent among these treatments were boar contact, transport stress and exogenous oestrogen administration. Furthermore, with the possible exception of oestrogen treatment (Hughes and Cole, 1978), this synchronization tended to continue up to the second and even third oestrus. Hence transport stress and, better still, boar contact prior to puberty can offer an alternative method of synchronizing the mating oestrus.

In the sow the problem of oestrus synchronization may be resolved to a large extent by a simple management procedure. Since sows generally return to heat 4–5 days after weaning (with the exception of very early-weaned sows – *see* Chapter 9) it would seem that weaning sows in groups should afford a high degree of synchronization of the postweaning oestrus. Oestrus synchronization obtained by batch weaning is as good as that observed when oestrus suppressants or stimulants are used, with the added advantage that subsequent fertility is not adversely influenced.

Thus, although attempts to artificially control the oestrous cycle of the pig have met with little success, alternatives are available for the synchronization of the mating oestrus. In the gilt the pubertal oestrus may be synchronized, and this should result in subsequent heats occurring with a high degree of synchronization. In the sow the practice of group weaning affords good synchronization of the post-weaning oestrus. Until such time as a practical system for the artificial control of the pig's oestrous cycle becomes available, reasonable alternatives are therefore at hand.

4.6 References

ANDERSON, L.L. (1966). *J. Reprod. Fert., Suppl.* **1**, 21

ANDERSON, L.L. and MELAMPY, R.M. (1967). In *Reproduction in the Female Mammal* (Ed. by G.E. Lamming and E.C. Amoroso). London, Butterworths

BAJEZ, E. (1952). *Anim. Breed. Abstr.* **20**, 161

BAKER, L.N., ULBERG, L.C., GRUMMER, R.H. and CASIDA, L.E. (1954). *J. Anim. Sci.* **13**, 648

BAKER, R.D., SHAW, G.A. and DODDS, J.S. (1970). *Can. J. Anim. Sci.* **50**, 25

BAKER, T.G. (1973). In *Reproduction in Mammals. 1. Germ Cells and Fertilization* (Ed. by C.R. Austin and R.V. Short). Cambridge, Cambridge University Press

BENESCH, F. (1953). *Anim. Breed. Abstr.* **21**, 283

BRINKLEY, H.J., WILFINGER, W.W. and YOUNG, E.P. (1973). *J. Anim. Sci.* **37**, 303

CASIDA, L.E. (1965). *Anat. Rec.* **61**, 389

CHRISTENSEN, R.K., POPE, C.E., ZIMMERMAN-POPE, V.A. and DAY, B.N. (1973). *J. Anim. Sci.* **36**, 914

DAY, B.N., ANDERSON, L.L., HAZEL, L.N. and MELAMPY, R.M. (1959). *J. Anim. Sci.* **18**, 909

DAY, B.N. and POLGE, C. (1968). *J. Reprod. Fert.* **17**, 227

DIEHL, J.R. and DAY, B.N. (1974). *J. Anim. Sci.* **39**, 392

DU MESNIL DU BUISSON, F. (1966). Thesis, University of Paris

DZUIK, P.J. (1960). *J. Anim. Sci.* **19**, 1319

DZUIK, P.J., POLGE, C. and ROWSON, L.E.A. (1964). *J. Anim. Sci.* **23**, 37

EDQVIST, L.E. and LAMM, A.M. (1971). *J. Reprod. Fert.* **25**, 447

FIRST, N.L., STRATMAN, F.W., RIGOR, E.M. and CASIDA, L.E. (1963). *J. Anim. Sci.* **22**, 66

GARBERS, D.L. and FIRST, N.L. (1969). *J. Reprod. Fert.* **20**, 45

GARDNER, M.L., FIRST, N.L. and CASIDA, L.E. (1963). *J. Anim. Sci.* **22**, 132

GERRITS, R.J., FAHMING, M.L., MEADE, R.J. and GRAHAM, E.F. (1963). *J. Anim. Sci.* **21**, 1022

GERRITS, R.J. and JOHNSON, L.A. (1965). *J. Anim. Sci.* **24**, 917

GIBSON, E.W., JAFFE, S.C., LASLEY, J.F. and DAY, B.N. (1963). *J. Anim. Sci.* **22**, 858

GORDON, I. (1975). *Proc. Br. Soc. Anim. Prod.* **4**, 79

GROVES, T.W. (1967). *Vet. Rec.* **80**, 470

GUTHRIE, H.D. (1975). *Theriogenology* **4**, 69

GUTHRIE, H.D., HENRICKS, D.M. and HANDLIN, D.L. (1972). *Endocrinology* **91**, 675

GUTHRIE, H.D. and POLGE, C. (1976a). *J. Reprod. Fert.* **48**, 423

GUTHRIE, H.D. and POLGE, C. (1976b). *J. Reprod. Fert.* **48**, 427

HALLFORD, D.M., WETTEMAN, R.P., TURMAN, E.J. and OMTVEDT, I.T. (1975). *J. Anim. Sci.* **41**, 1706

HAMMOND, J. Jr., MASON, I.L. and ROBINSON, T.J. (1971). *Hammond's Farm Animals*. London, Arnold

HENRICKS, D.M., GUTHRIE, M.D. and HANDLIN, D.L. (1972). *Biol. Reprod.* **6**, 210

HUGHES, P.E. (1976). PhD Thesis, University of Nottingham

HUGHES, P.E. and COLE, D.J.A. (1978). *Anim. Prod.* **27**, 12

HUNTER, R.H.F. (1966). *Anim. Prod.* **8**, 457

HUNTER, R.H.F. (1967). *J. Reprod. Fert.* **13**, 133

HUNTER, R.H.F. and POLGE, C. (1966). *J. Reprod. Fert.* **12**, 525

KIDDER, H.E., CASIDA, L.E. and GRUMMER, R.H. (1955). *J. Anim. Sci.* **14**, 470

KMENT, A. and HALAMA, A. (1953). *Anim. Breed. Abstr.* **21**, 285

LAMMING, G.E., HAFS, H.D. and MANNS, J.G. (1975). *Proc. Br. Soc. Anim. Prod.* **4**, 79

LONGNECKER, D.E. and DAY, B.N. (1968). *J. Anim. Sci.* **27**, 709

NALBANDOV, A.V. (1964). *Reproductive Physiology,* 2nd edn. London, W.H. Freeman

NELLOR, J.E. (1960). *J. Anim. Sci.* **19**, 412

NELLOR, J.E., AHRENHOLD, J.E., FIRST, N.L. and HOEFER, J.A. (1961). *J. Anim. Sci.* **20**, 22

POLGE, C. (1972). In *Pig Production* (Ed. by D.J.A. Cole). London, Butterworths

POND, W.G., HANSEL, W., DUNN, J.A., BRATTON, R.W. and FOOTE, R.H. (1965). *J. Anim. Sci.* **24**, 536

RAYFORD, P.L., BRINKLEY, H.J. and YOUNG, E.P. (1971). *Endocrinology* **88**, 707

RICHARDS, J.S., RAO, M.C. and IRELAND, J.J. (1978). In *Control of Ovulation* (Ed. by D.B. Crighton, N.B. Haynes, G.R. Foxcroft and G.E. Lamming). London, Butterworths

SCHILLING, E. and CERNE, F. (1972). *Vet. Rec.* **71**, 471

SHEARER, I.J., PURVIS, K., JENKIN, G. and HAYNES, N.E. (1972). *J. Reprod. Fert.* **30**, 347

SIGNORET, J.P. (1972). In *Pig Production* (Ed. by D.J.A. Cole). London, Butterworths

SPORRI, H. and CANDINAS, L. (1951). *Anim. Breed. Abstr.* **19**, 1833

STABENFELDT, G.H., ATKINS, E.L., EWING, L.L. and MORRISETTE, M.C. (1969). *J. Reprod. Fert.* **20**, 443

STIFT, K. (1953). *Anim. Breed. Abstr.* **21**, 66

STRATMAN, F.W. and FIRST, N.L. (1965). *J. Anim. Sci.* **24**, 930

TANABE, T.Y., WARNICK, A.C., CASIDA, L.E. and GRUMMER, R.H. (1949). *J. Anim. Sci.* **8**, 550

ULBERG, L.C., GRUMMER, R.H. and CASIDA, L.E. (1951). *J. Anim. Sci.* **10**, 665

The influence of parity, or number of previous litters, on ovulation rate is also well established (Hammond, 1914; Burger, 1952; Perry, 1954; King and Young, 1957). Indeed, the work of Perry (1954) clearly demonstrates that ovulation rate shows a marked increase over the first four parities, reaching a plateau at about the

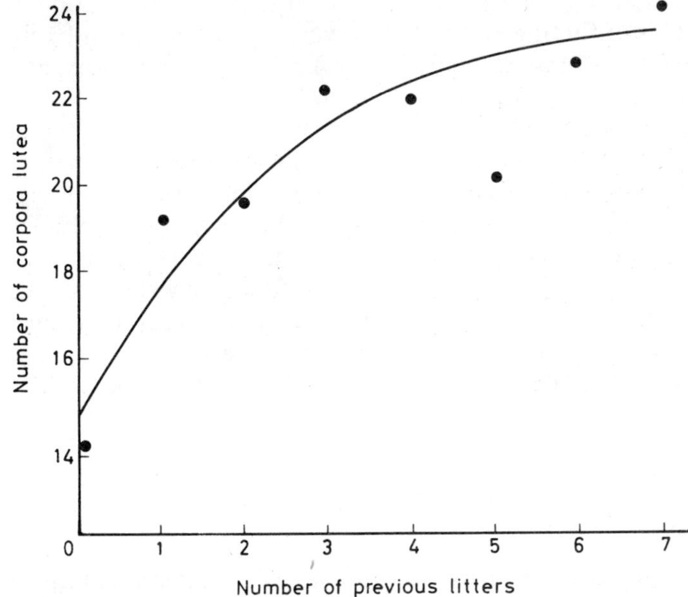

Figure 5.1 The effect of parity number on ovulation rate in the pig (from Perry, 1954)

sixth parity (*Figure 5.1*). There is no apparent drop in ovulation rate later in the sow's life, although litter size is likely to be reduced as a result of an increase in embryo mortality (*see* Chapter 7).

WEIGHT AND BODY CONDITION

Since increases in age and parity are usually associated with increases in body weight, it might be expected that ovulation rate would also be correlated with the weight of the animal. However, body weight is a manifestation of many interacting factors such as age, breed, nutrition and disease, and thus the association with ovulation rate is not a simple one. In addition, any observed influence of weight on ovulation rate may be due to either absolute body weight or the dynamic effect of weight change.

There is much conflicting evidence concerning the influence of absolute weight on ovulation rate. In many experiments no relationship has been found between the two, in either the gilt (Zimmerman *et al.*, 1960; Kirkpatrick *et al.*, 1967) or the sow (King and Young,

1957; Hardy and Lodge, 1969). However, Heap, Lodge and Lamming (1967) have reported that there is an increase in ovulation rate of 0.73 ova for every 10 kg increase in service weight of the sow. A similar relationship has also been reported by other workers (Bowman, Bowland and Fredeen, 1961; Omtvedt, Stanislav and Whatley, 1965; O'Bannon *et al.*, 1966). Clearly it is difficult to draw conclusions from such conflicting results. However, it does not seem likely that body weight *per se* is very closely correlated with ovulation rate, although it may be in the case of dynamic weight change.

Pike and Boaz (1967) observed that ovulation rate was increased when sows were fed a high plane diet (3.6 kg/day) prior to mating compared to low plane feeding (1.8 kg/day). Since this effect was only apparent in 'thin' sows, it seems likely that it is attributable to increasing body weight and condition. Several more recent reports also suggest that a correlation exists between dynamic weight change and ovulation (Dyck, 1974; Young, Omtvedt and Johnson, 1974). However, it is possible that these results may be a reflection of the 'flushing effect' (*see below*) rather than a true effect of weight change.

NUTRITION

The effects of nutrition on ovulation rate are of considerable importance from a practical viewpoint. They have been the subject of much research effort and comprehensive reviews have been made by Brooks (1970) and Anderson and Melampy (1972). The more important aspects of this relationship are discussed below.

First it is necessary to define which component of the diet is associated with observed variations in ovulation rate. The present evidence indicates that the level of energy received by the female pig is the dietary factor primarily responsible for alterations in ovulation rate. However, it is worth noting that, under certain conditions, protein level may also exert some influence. Increasing protein content of the diet has been reported to cause slight increases in ovulation rate (Adams *et al.*, 1960; Corman and Zimmerman, 1972), whereas protein-free diets result in decreased ovulation rates after 4 to 6 cycles (Adams *et al.*, 1960: McGillivray *et al.*, 1964). Under most nutritional circumstances, however, the influence of protein level on ovulation rate may be considered minimal.

The effects of dietary energy on ovulation rate may be considered in two categories; long term influences which relate to prepubertal, oestrous cycle and weaning to remating nutrition, and short term effects resulting from manipulation of dietary energy content at, or around, the time of oestrus.

Long-term effects

As early as 1951 Robertson *et al.* (1951a) noted that gilts on *ad libitum* feeding had higher ovulation rates at their first and second heats than similar gilts which were fed approximately 70 per cent of the *ad libitum* food intake from age 70 days onwards. Subsequently, Self, Grummer and Casida (1955) carried out a similar study in which *ad libitum* feeding (HIGH) was compared with 66 per cent of *ad libitum* feed intale (LOW). In this experiment, however, half of the gilts in each treatment group were changed over at first oestrus to the alternative plane of nutrition, hence giving four treatment groups in all (HIGH–HIGH, HIGH–LOW, LOW–HIGH, LOW–LOW). The results, summarized in *Table 5.5*, demonstrate that high plane feeding

Table 5.5. The effect of feed level prior to puberty and during the first oestrous cycle on ovulation rate in the gilt

Feeding regimen	*Ovulation rate**
HIGH-HIGH	13.9
HIGH-LOW	11.1
LOW-HIGH	13.6
LOW-LOW	11.1

From Self, Grummer and Casida (1955).
*Ovulation rates given refer to the number of corpora lutea present following second oestrus.

for only three weeks (between first and second heat) was sufficient to produce the improved ovulatory response. These results have since been verified by many workers, and a comprehensive list of data has been presented by Anderson and Melampy (1972). Indeed, it can now be stated that gilts and sows consistently show a positive response in ovulation rate to high plane feeding.

However, what is the effect on ovulation rate if the period of high plane feeding is reduced below 21 days? In a review of 39 experiments, Anderson and Melampy (1972) reached the conclusion that

Table 5.6. The effect of high energy intake on ovulation rate in the pig

Number of trials	*Days pigs given high energy diet before oestrus*	*Increase in ovulation rate*
6	0–1	1.35
6	2–7	0.86
8	10	1.58
14	11–14	2.23
5	17–21	0.66

From Anderson and Melampy (1972).

the optimum duration of high energy feeding prior to oestrus was 11–14 days (*Table 5.6*). However, it is worth noting that the original data relating to 17–21 days 'flushing' (a term used to describe this system of high plane feeding prior to oestrus) incorporates three sets of data from one report (Zimmerman, 1966) which clearly do not

Table 5.7. The effect of high energy intake for 17–21 days prior to oestrus on ovulation rate in the pig

Source	*Days diets fed prior to oestrus*	*Metabolizable energy intake* (MJ/day)		*Ovulation rate*		
		Restricted diet	*Fixed diet*	*Restricted diet*	*Full diet*	*Difference*
Zimmerman (1966)	17	30.0	37.1	16.5	16.6	+0.1
Zimmerman (1966)	17	30.0	37.1	16.0	15.0	−1.0
Zimmerman (1966)	17	30.0	37.1	15.9	15.0	−1.9
Short *et al.* (1963)	21	–	–	14.8	17.7	+2.9
McGillivray *et al.* (1963)	21	22.6	45.9	14.8	18.0	+3.2
				Average increase in ovulation rate	=	+0.66*

*This is the figure quoted by Anderson and Melampy (1972) in *Table 5.6*.

conform to the overall pattern. This could be due to small differences between high and low energy intakes, and the relatively low level used in the high energy diet (*Table 5.7*).

Short term nutrition

The data presented in *Table 5.6* indicate that increases in ovulation rate may be obtained by high energy feeding for just one day at the time of oestrus. Such short term feeding influences were the subject of a series of experiments by Brooks (1970). In this work it was found that increasing the feed level from 1.8 to 3.6 kg/day for one feed on the day following first mating (mating twice on consecutive days at the appearance of oestrus) had no effect on the ovulation rate of gilts mated at third heat. However, when feed level was increased from 1.8 to 3.6 kg/day for one feed on the first day of mating it was found that ovulation rate was raised significantly from 11.9 to 13.2. This finding was then further investigated in a co-ordinated field trial using 16 centres and 581 animals. Sows were fed either 2.7 kg/day or 5.4 kg/day on the first day of mating. Overall the results showed no effect of this change in feed intake on litter size, and only one centre observed any significant increase in the number of piglets born per litter (from 10.7 to 12.0 piglets).

In conclusion, it appears two principle factors influence the 'flushing' effect: the timing of the increased feed intake and the basal level of feeding prior to this. It seems likely that short-term flushing will have little effect on the sow as ovulation rate is normally in excess of requirements. However, in some circumstances this form of flushing may have beneficial effects on the ovulation rate, and hence reproductive performance, of gilts. Such circumstances may only arise when basal feeding has been below what is normally acceptable in commercial practice.

CLIMATIC ENVIRONMENT

Climatic influences on reproductive performance in the pig have, until recently, only been considered in terms of season of birth or farrowing. However, more recent work has split seasonal effects into the two main contributing factors, namely ambient temperature and daylength. Therefore, the influence of the climatic environment on ovulation rate is considered here in the three categories of season, ambient temperature and daylength.

The effects of season on ovulation rate are by no means clear. In two experiments Gossett and Sorensen (1959) and Sorensen, Thomas and Gossett (1961) observed contradictory effects of season on ovulation rate (*Table 5.8*). However, most other workers have

Table 5.8. The effect of season on ovulation rate in the pig

Source	*Litter size*	
	Autumn farrowing	*Spring farrowing*
Gossett and Sorensen (1959)	10.9	13.0
Sorensen *et al*. (1961)	12.3	11.0

been unable to show a seasonal effect on either ovulation rate (Moore *et al*., 1973) or litter size (Braude, Clark and Mitchell, 1954; Ahlshwede and Robinson, 1966).

The influence of ambient temperature on ovulation rate was studied in a recent experiment by Teague, Roller and Griffo (1968). These workers investigated the effects of three temperatures (26.7, 30.0 and 33.3 °C) applied over one oestrous cycle prior to breeding, on ovulation rate and conception rate. They observed that ovulation rate reduced significantly as ambient temperature increased, with ovulation rates of 14.2, 13.7 and 13.1 respectively. There was also a significant increase in the number of gilts returning to oestrus as the ambient temperature increased. Since a rise in temperature of 6.6 °C resulted in a 45 per cent reduction in food intake in this experiment, the observed results are likely to be due to energy intake rather than ambient temperature *per se*.

The effects of light on ovulation rate in the pig have been studied in a number of experiments, utilizing lighting regimes ranging from complete darkness to continuous illumination. In all this work no significant effect of daylength on ovulation rate has been reported (Dufour and Bernard, 1968; Waddill, Chaney and Butt, 1968; Hacker, King and Bearss, 1974). However, recent reports from Russia (Belyaev, Klotchkov and Klotchkova, 1969) and the USA (Hacker, King and Bearss, 1974) have suggested that daylength may influence litter size through an effect on embryo survival rather than ovulation rate.

EXOGENOUS HORMONES

The administration of exogenous hormones as a means of inducing higher ovulation rates in the pig has been the subject of much work in recent years. The predominant hormone preparation used is a combination of PMS and HCG. This is normally given during the follicular phase of the oestrous cycle in order to promote the development of more follicles to ovulatory size at the time of ovulation. When administered during the luteal phase of the cycle PMS/HCG preparations are not effective, and often result in adverse side effects (Tanabe *et al.*, 1949; Spalding, Berry and Moffitt, 1955; Hunter, 1966). The optimum timing of application for gilts appears to be day 15 or 16 of the oestrous cycle (Tanabe *et al.*, 1949; Hunter, 1964). Indeed, Hunter (1964) found that injections of PMS on day 14 of the cycle prolonged the cycle to 28–29 days, and Longnecker, Lasley and Day (1965) found little response when treatment was applied on day 17. For sows, the optimum treatment time appears to be on the day of weaning (Tanabe *et al.*, 1949; Longnecker and Day, 1968).

The use of PMS alone, or a combination of PMS and HCG during the follicular phase of the cycle has been clearly demonstrated to increase ovulation rate (*Table 5.9*). The number of ovulations follows a typical dose response relationship to the amount of hormone injected (Baker and Coggins, 1966, 1968). Indeed, Hunter (1964) reported that, above a dose of 500 i.u. of PMS, ovulation rate was raised by 1.9 additional eggs shed for each 100 i.u. increment of PMS administration.

In contrast, FSH application has not been as effective as PMS treatment in either single or repeated doses (Spalding, Berry and Moffitt, 1955; Day, Romack and Lasley, 1961). The difference

Table 5.9. The effect of exogenous hormones administered during the follicular phase of the oestrous cycle on ovulation rate in the pig

Source	*Treatment*		*Ovulation rate*	
	PMS (i.u.)	*HCG* (i.u.)	*Control*	*Treated*
Tanabe *et al.* (1949)	2000	sheep pituitary extract	–	26.5
Gibson *et al.* (1963)	1200	500	10.7	23.8
Hunter (1964)	500	–	13.3	15.0
	750	–	13.3	24.0
	1000	–	13.3	24.3
	1250	–	13.3	25.3
	1500	–	13.3	38.5
Phillipo (1968)	1000	850	11.7	15.5
	1250	850	11.7	22.5
	1500	850	11.7	20.3

between the two preparations may be attributed either to the much longer half life of PMS in the circulation (Cole, 1969), or to the fact that PMS has both FSH and LH properties which may provide a better balance for the stimulation of normal follicular growth than FSH alone.

Although ovulation rate can clearly be increased by the administration of exogenous hormone preparations, many workers have reported that litter size following such treatments is not significantly increased above that of controls (Day, Romack and Lasley, 1961; Longnecker, Lasley and Day, 1965; Wood *et al.*, 1967). However, Deneke and Day (1973) have observed significant increases in litter size on day 70 of gestation following treatment with PMS. Furthermore Longnecker and Day (1968) found that litter size was increased by two piglets over controls when sows had been treated with 1200 i.u. of PMS prior to the first postweaning oestrus. These variations in reaction to superovulation are probably due to such factors as the inability of the uterus to maintain a large number of embryos, and the induction of ovarian secretion by the administered hormones.

5.2 Optimization of ovulation rate

It has previously been mentioned that any limitation to litter size due to low ovulation rate is more likely to occur in the gilt than in the sow. In nearly all cases, in fact, the sow ovulates more eggs than she is capable of maintaining as viable embryos through to parturition. However, it would seem reasonable to assume that ovulation rate may be limiting in the weaned sow and therefore to take some measure to avoid this. The normal form that this may take is the generous allowance of feed from weaning up to and including remating, in order to provide a flushing effect. The cost of such a measure is minimal, since the weaning to remating interval is a relatively short period within the reproductive cycle of the pig.

In contrast, ovulation rate may be a primary factor limiting litter size in the gilt. This will almost certainly be the case if ovulation rate is below 12–14 ova. This target ovulation rate of 12–14 ova is often given as an arbitrary optimum for the gilt, since the size of litter the gilt is capable of producing is restricted to 11–12 pigs by 'maternal limitation' (*see* Chapter 7). To achieve this rate of ovulation one of several methods may be employed. First, nature may be allowed to take its course and normal sexual development, and the concommitant rise in ovulation rate, awaited. This would involve allowing the gilt to attain an advanced stage of development (in terms of age and liveweight), and ensuring that she had experienced two or three oestrous cycles before first mating. However, this results in a long period of non-productivity prior to first mating and thus causes a rise in the cost of the replacement gilt. Secondly, exogenous hormones (e.g. PMS/HCG) may be administered in order to stimulate the release of more ova. This constitutes a viable method of boosting ovulation rate but does contain the inherent risk of superovulation, resulting in a reduction in litter size. A third method of raising ovulation rate is to employ the technique of flushing once again. In the case of the gilt this appears to be most effective (and practical) when applied over the course of one oestrous cycle prior to mating (*Table 5.5*). Using this technique the number of oestrous cycles that the gilt needs to experience prior to breeding (in order to raise ovulation rate to the required level) may be reduced. Furthermore, this system not only results in saving time, it also reduces the cost of feeding the replacement gilt since the nutrient requirements over one or two oestrous cycles are saved at the expense of a higher feed level for just one oestrous cycle.

5.3 Conclusions

It is clear that ovulation rate may limit the size of litter produced by the gilt. A minimum ovulation rate of 12–14 ova is suggested for the gilt, since this is above the litter upper limit of 11–12 pigs due to 'maternal limitation'. This ovulation rate may be achieved by flushing the gilt for one oestrous cycle prior to mating, where a level of 3–4 kg feed/day (of a conventional sow ration providing approximately 13

MJ of ME/kg and 14 per cent crude protein) is considered adequate. Furthermore, this flushing may be introduced between pubertal and second oestrus, since this should result in an ovulation rate within the range 12–14 ova. Thus, mating may take place during the second oestrous period, this reducing rearing cost and time.

Flushing is also recommended for the sow over the short period from weaning until remating. This may not be necessary for the majority of sows, since ovulation rate will be in excess of requirements, but is a good form of insurance against the occasional sow whose ovulation rate may be low in the absence of flushing. In addition, the sow at weaning does tend to be in low condition and thus extra feeding in this period may help restore condition and stimulate return to oestrus. Finally, the level of feeding envisaged for the sow over this period is approximately 3–4 kg/day.

5.4 References

ADAMS, C.R., BECKER, D.E., TERRILL, S.W., NORTON, H.W. and JENSEN, A.H. (1960). *J. Anim. Sci.* **19**, 1245 (abstract)

AHLSCHWEDE, W.T. and ROBINSON, O.W. (1966). *J. Anim. Sci.* **25**, 916 (abstract)

ANDERSON, L.L. and MELAMPY, R.M. (1972). In *Pig Production* (Ed. by D.J.A. Cole). London, Butterworths

BAKER, L.N., CHAPMAN, A.B., GRUMMER, R.H. and CASIDA, L.E. (1958). *J. Anim. Sci.* **17**, 612

BAKER, R.D. and COGGINS, E.G. (1966). *J. Anim. Sci.* **25**, 918 (abstract)

BAKER, R.D. and COGGINS, E.G. (1968). *J. Anim. Sci.* **27**, 1607

BELYAEV, D.K., KLOTCHKOV, D.V. and KLOTCHKOVA, A.J. (1969). *Anim. Breed. Abstr.* **38**, 3991

BHALLA, R.C., FIRST, N.L., CHAPMAN, A.B. and CASIDA, L.E. (1969). *J. Anim. Sci.* **28**, 780

BOWMAN, G.H., BOWLAND, J.P. and FREDEEN, H.T. (1961). *Can. J. Anim. Sci.* **41**, 220

BRAUDE, R., CLARK, P.M. and MITCHELL, K.G. (1954). *J. agric. Sci., Camb.* **45**, 19

BROOKS, P.H. (1970). PhD Thesis, University of Nottingham

BURGER, J.P. (1952). *Onderstepoort J. vet. Res. Suppl.* **2**, 1

COLE, H.N. (1969). In *Reproduction in Domestic Animals* (Ed. by H.H. Cole and P.T. CUPPS) 2nd edn. New York, Academic Press

CORMAN, R. and ZIMMERMAN, D.R. (1972). *J. Anim. Sci.* **35**, 1103 (abstract)

CUNNINGHAM, P.J. and ZIMMERMAN, D.R. (1973). *J. Anim. Sci.* **37**, 231 (abstract)

DAY, B.N., ROMACK, F.E. and LASLEY, J.F. (1961). *J. Anim. Sci.* **20**, 969 (abstract)

DENEKE, W.A. and DAY, B.N. (1973). *J. Anim. Sci.* **36**, 1137

DUFOUR, J. and BERNARD, C. (1968). *Can. J. Anim. Sci.* **48**, 425

DYCK, G.W. (1974). *Can. J. Anim. Sci.* **54**, 287

GIBSON, E.W., JAFFE, S.C., LASLEY, J.F. and DAY, B.N. (1963). *J. Anim. Sci.* **22**, 858 (abstract)

GOODE, L., WARNICK, A.C. and WALLACE, H.D. (1965). *J. Anim. Sci.* **24**, 959

GOSSETT, J.W. and SORENSEN, A.M. Jr. (1959). *J. Anim. Sci.* **18**, 40

HACKER, R.R., KING, G.J. and BEARSS, W.H. (1974). *J. Anim. Sci.* **39**, 155 (abstract)

HAINES, C.E., WARNICK, A.C. and WALLACE, H.D. (1959). *J. Anim. Sci.* **18**, 347

HAMMOND, J. (1914). *J. agric. Sci., Camb.* **6**, 263

HARDY, B. and LODGE, G.A. (1969). *J. Reprod. Fert.* **19**, 555

HAUSER, E.R. DICKERSON, G.E. and MAYER, D.T. (1952). *Res. Bull. Mo. agric. Exp. Stn* **503**

HEAP, F.C., LODGE, G.A. and LAMMING, G.E. (1967). *J. Reprod. Fert.* **13**, 269

HUGHES, P.E. (1976). PhD Thesis, University of Nottingham

HUNTER, R.H.F. (1964). *Anim. Prod.* **6**, 189

HUNTER, R.H.F. (1966). *Anim. Prod.* **8**, 457

KING, J.W.R. and YOUNG, G.B. (1957). *J. agric. Sci., Camb.* **48**, 457

KIRKPATRICK, R.L., HOWLAND, B.E., FIRST, N.L. and CASIDA, L.E. (1967). *J. Anim. Sci.* **26**, 188

LASLEY, E.L. (1957). *J. Anim. Sci.* **16**, 335

LERNER, E.H., MAYER, D.T. and LASLEY, J.F. (1957). *Res. Bull. Mo. agric. Exp. Stn* **629**

LONGNECKER, D.E. and DAY, B.N. (1968). *J. Anim. Sci.* **27**, 709

LONGNECKER, D.E., LASLEY, J.F. and DAY, B.N. (1965). *J. Anim. Sci.* **24**, 924 (abstract)

MACPHERSON, R.M., JONES, A.S. and HOVELL, F.D. de B. (1973). *Proc. Br. Soc. Anim. Prod.* 2, 88 (abstract)

McGILLIVRAY, J.J., NALBANDOV, A.V., JENSEN, A.H., NORTON, H.W., HARMON, G.B. and BECKER, D.E. (1963). *J. Anim. Sci.* 22, 1127 (abstract)

McGILLIVRAY, J.J., NALBANDOV, A.V., JENSEN, A.H., NORTON, H.W., HARMON, B.G. and BECKER, D.E. (1964). *J. Anim. Sci.* **23**, 1214 (abstract)

MEAT and LIVESTOCK COMMISSION (M.L.C.). (1976). *Feed Recording Services Annual Report*

MOORE, C.P., DUTT, R.H., HAYS, V.W. and CROMWELL, G.L. (1973). *J. Anim. Sci.* **37**, 734

NEWMAN, J.A. (1963). *Can. J. Anim. Sci.* **43**, 285

O'BANNON, R.H., WALLACE, H.D., WARNICK, A.C. and COMBS, G.E. (1966). *J. Anim. Sci.* **25**, 706

OMTVEDT, I.T., STANISLAW, C.M. and WHATLEY, J.A. Jr. (1965). *J. Anim. Sci.* **24**, 531

PERRY, J.S. (1954). *J. Embryol. exp. Morph.* **2**, 308

PHILLIPPO, M. (1968). *Adv. Reprod. Physiol.* **3**, 148

PIKE, I.H. and BOAZ, T.G. (1967). In *Proceedings of a symposium on nutrition of sows, University of Nottingham, 1967*, pp. 13–15. London, PIDA

RATHNASABAPATHY, V., LASLEY, J.F. and MAYER, D.T. (1956). *Res. Bull. Mo. agric. Exp. Stn* **615**

REDDY, V.B., LASLEY, J.F. and MAYER, D.T. (1958). *Res. Bull. Mo. agric. Exp. Stn* **666**

ROBERTSON, G.L., CASIDA, L.E., GRUMMER, R.H. and CHAPMAN, A.B. (1951a). *J. Anim. Sci.* **10**, 841

ROBERTSON, G.L., GRUMMER, R.H., CASIDA, L.E. and CHAPMAN, A.B. (1951b). *J. Anim. Sci.* **10**, 647

SELF, H.L., GRUMMER, R.H. and CASIDA, L.E. (1955). *J. Anim. Sci.* **14**, 573

SHORT, R.E., ZIMMERMAN, D.R. and SUMPTION, L.J. (1963). *J. Anim. Sci.* 22, 868 (abstract)

SORENSEN, A.M. Jr., THOMAS, W.B. and GOSSETT, J.W. (1961). *J. Anim. Sci.* **20**, 347

SPALDING, J.F., BERRY, R.O. and MOFFITT, J.G. (1955). *J. Anim. Sci.* **14**, 609

SQUIERS, C.D., DICKERSON, G.E. and MAYER, D.T. (1952). *J. Anim. Sci.* **9**, 683 (abstract)

STEWART, H.A. (1945). *J. Anim. Sci.* **4**, 250

TANABE, T.Y., WARNICK, A.C., CASIDA, L.E. and GRUMMER, R.H. (1949). *J. Anim. Sci.* **8**, 550

TEAGUE, H.S., ROLLER, W.L. and GRIFFO, A.P. Jr. (1968). *J. Anim. Sci.* **27**, 408

WADDILL, D.G., CHANEY, C.H. and DUTT, R.H. (1968). *J. Reprod. Fert.* **15**, 123

WARNICK, A.C., WIGGINS, E.L., CASIDA, L.E., GRUMMER, R.H. and CHAPMAN, A.B. (1951). *J. Anim. Sci.* **10**, 479

WARREN, W.M. and DICKERSON, G.E. (1952). *Res. Bull. Mo. agric. Exp. Stn* **511**

WEBEL, S.K., PETERS, J.B. and ANDERSON, L.L. (1970). *J. Anim. Sci.* **30**, 565

WOOD, R.D., CHANEY, C.H., WADDILL, D.G. and DUTT, R.H. (1967). *J. Anim. Sci.* **26**, 231 (abstract)

YOUNG, L.D., OMTVEDT, I.T. and JOHNSON, R.H. (1974). *J. Anim. Sci.* **39**, 480

ZIMMERMAN, D.R. (1966). *J. Anim. Sci.* **25**, 1268 (abstract)

ZIMMERMAN, D.R., SPIES, H.G., RIGOR, E.M., SELF, H.L. and CASIDA, L.E. (1960). *J. Anim. Sci.* **19**, 687

Chapter 6

Fertilization and conception

This chapter follows the egg shed at ovulation through to the first cleavage divisions following successful fertilization. Clearly this is a short but very critical stage in the reproductive process and although the pig in a good environment can achieve high levels of reproductive efficiency in terms of fertilization rate and conception rate relative to other species, there are a multiplicity of factors which can influence the level reached in practice.

Fertilization initiates gestation and its success depends largely on the timing of insemination (Haring, 1937; Dzuik, 1970). Under normal conditions fertilization rate (or the percentage of viable ova shed at ovulation which are fertilized and begin cleavage division) is between 90 and 100 per cent (Casida, 1953; Hancock, 1958; Perry and Rowlands, 1962; Self, Grummer and Casida, 1955; Squiers, Dickerson and Mayer, 1951).

The percentage of sows farrowing to the first post weaning service (farrowing rate) gives a quantitiative indication of the success of conception. More commonly the conception rate refers to the percentage of sows which appear to hold to this first service as indicated by the number of sows which do not return to oestrus 3 weeks later. This latter concept can be misleading because a significant proportion of sows appear to hold to the first service and do not show a heat 3 weeks later but are found non-pregnant much later. These sows therefore did not achieve successful conception at the first service. Farrowing rates are usually high and of the order of 80–90 per cent (Aumaitre, 1972; te Brake, 1972; Van der Heyde, 1972; Scofield and Penny, 1969) but much variability exists.

6.1 Physiology of fertilization and conception

SPERM TRANSPORT

The eggs shed at ovulation are gathered by the finger-like fimbriae of the ovarian end of the oviduct and if capacitated (mature) sperm cells are present in this part of the female's reproductive tract then fertilization should promptly be effected. The timing of the sequence of events leading up to the fusion of the nuclear material of both sperm and egg is crucial to the success of the process. In particular, the timely passage of sperm to the ampullae of the oviduct from the site of deposition of semen, which is the cervical end of the uterus, must be effected. With poor sperm transport linked with incorrectly timed insemination, the ova will in all probability degenerate and be expelled before mature sperm cells can encounter them.

It was once thought that the passage of sperm cells up the female's reproductive tract was a function of the motility of the sperm or the ability of the whip-like flagella of sperm cells to propell the gamete to its destination. It can no longer be doubted that the major causal factors in sperm transport are the uterine and oviductal contractions. Observations on the ovarian end of the oviduct shortly after mating have shown that dead sperm cells and other particles such as Indian ink are transported with equal ease and speed as motile sperm cells. Under normal circumstances, following mating some sperm cells arrive in the ampullae within minutes of insemination but it seems unlikely that these early arrivers can have immediate fertilizing capacity because of the necessary maturation phase (capacitation) which must occur to the sperm cells in the ampullae before they acquire the ability to fertilize eggs. Hence sperm must be deposited in the uterus at mating or insemination some hours prior to ovulation to allow time for both capacitation and sperm transport.

UTERINE TRANSPORT OF SPERM

Copulation in pigs results in the deposition in the uterus, via a relaxed cervical canal, of a voluminous quantity of semen of low sperm concentration. Hence in the pig the cervix cannot provide the filtration and selection mechanism on sperm cells observed in the sheep and cow (Moghissi, 1972; Hafez, 1973). Between 30 and 60 billion sperm cells begin the ascent to the ampullae but only a few hundred actually reach the site of fertilization. A great deal of culling out of sperm cells therefore must occur in both the uterus and oviduct. It has been postulated that this is a natural selection mechanism enabling the most viable sperm with the most viable potential genotype to be the ones which produce the new offspring. There is at present scant evidence concerning the pattern of sperm migration in the uterine lumen. Some spermatozoa appear to enter the endometrial glands where they are presumably destroyed, while others are destroyed by phagocytosis carried out by leucocytes which exist in the lumen of the uterus. The relative losses of spermatozoa as a result of these processes is however unknown.

It is the myometrium which plays a major role in the transport of sperm in the uterus. This effect is initiated by the onset of oestrus causing release of small amounts of oxytocin from the posterior pituitary gland. Oxytocin causes rhythmic contractions of the smooth muscle myometrial layer in the uterine wall and this is then augmented by the action of copulation. At mating, therefore, the uterine wall is undergoing strong rhythmic contractions under the influence of oxytocin. This results in any sperm cells appearing in the uterus being rapidly transported to the oviducts.

SPERM TRANSPORT IN THE OVIDUCTS

In the pig the bulk of sperm cells deposited at mating disappear from the uterus within 2 hours of copulation (Hafez, 1976) leaving a high concentration of sperm cells at the uterotubal junction. This pool of sperm cells persists for about 24 hours and then disappears within the following 48 hours. Thus a gradual and metered flow of sperm cells passes from the uterotubal junction up to the ovarian end of the oviduct. The junction between uterus and oviduct seems to play a controlling role in allowing only a steady flow of sperm cells into the oviducts.

The pattern and rate of sperm transport when actually in the oviduct is controlled by a number of mechanisms such as: peristalsis of oviductal musculature, contractions of the oviductal mucosal folds and mesosalpinx, fluid movements as a result of ciliary movement and in addition the opening and closing of the valve-like uterotubal junction. The relative contribution each one of these mechanisms makes to sperm movement is not known. It is probable however that, as well as these mechanisms promoting the movement of sperm through the oviduct, the junction between the isthmus and ampullae acts as a selective barrier to sperm cells only allowing a passage of a certain amount. In the absence of this junction, sperm cells still pass to the fimbriae at the same rate but an increased incidence of polyspermy has been observed (Hunter and Leglise, 1971). This demonstrates that although there is only a small anatomic difference between isthmus and ampullae, this makes in practice a significant contribution to quantitative selection of sperm cells reaching the fimbriae.

On reaching the fimbriae sperm cells may either fertilize any egg cells present (assuming capacitation has occurred) or they can remain in the oviduct for a time awaiting ovulation. Sperm cells may also be lost in a number of ways. First, sperm cells are lost by the leucocytic response: in fact phagocytosis occurs at a slightly reduced rate in the oviduct compared to the uterus. Secondly, damaged sperm cells may be carried back to the uterus, via the ciliated cells in the oviducts, and eventually expelled via the vagina. Thirdly, a significant proportion of sperm cells reaching the fimbriae is released into the peritoneal cavity. It has been noted in laboratory species (Rowlands, 1957) that sperm can traverse the peritoneal cavity and enter the fimbriae on the other tubal ending and still have the capacity to fertilize eggs in the other horn.

The whole process of transportation of sperm throughout both the uterus and oviduct is under neuro-hormonal control and as mentioned earlier oxytocin plays a direct causal role in uterine contractions. In addition to posterior pituitary involvement in sperm transport, ovarian hormones also affect the structure and ultrastructure and hence the secretory activity of the epithelial layer in both the uterus and oviduct. As a consequence of these changes the contractile abilities of the uterine wall and the oviduct wall are modified by the balance of circulating oestrogen and progosterone. Prostaglandins are also contained within the seminal plasma and it seems likely that this contributes to the waves of contraction in the smooth muscle of the reproductive tract.

The central nervous system in the female must play a coordinating function in the endocrine control of sperm transport and this presents the possibility that psychological stress or any psychosomatic factor may inhibit the timely passage of semen. It is known that adrenaline is released under stress and this has an antagonistic effect on oxytocin which would in turn inhibit the myometrium contractile response.

It is also known that sperm motility and transport are sensitive to the biochemical conditions prevailing in the female's tract. In particular the pH status of uterine and oviductal fluids mediates sperm activity and flagella function.

CAPACITATION

Spermatozoa are not immediately able to fertilize eggs upon deposition in the uterus and a period of about 2–4 hours must elapse for the sperm to be exposed to uterine and oviductal fluids before successful fertilization can be achieved. This process of final sperm maturation or capacitation has much bearing on the timing of insemination and in view of the short viable life span of the female ova the aim is to inseminate a number of hours prior to ovulation in order that capacitated spermatozoa are 'lying in wait' for egg cells as they are shed at ovulation.

Capacitation results in steady release of a number of hydrolytic enzymes from the sperm cell which are capable of assisting in the digestive passage of the spermatozoa through the outer membranes of the egg cell. Many factors have been suggested as participating agents in capacitation. These include: calcium ions, fertilizins and viruses although the precise role of each has yet to be ascertained.

The main event occurring during capacitation is probably the increase in the permeability of the cell membrane in the acrosomal region (Austin, 1969). This might be brought about by the removal of the inhibitory effect of certain compounds present in the seminal plasma including proteinase inhibitors and a decapacitation factor. In the epididymis the presence of these compounds permits metabolism and development of the gametes but renders them infertile while contact between cells and seminal plasma exists. Upon entering the female tract these inhibitory factors are gradually removed. This allows the spermatozoa to complete their last phase of development.

Capacitation is probably under hormonal control and oestrogen is known to stimulate the process and progesterone inhibits it.

FERTILIZATION

Once the process of capacitation is complete in the ampullae, if ovulation occurs then the phenomena associated with fertilization can commence. Of obvious importance in this mechanism is the initial encounter of sperm with ova. Probably only a thousand or so viable sperm cells actually reach the ovarian ends of the oviducts to fertilize the eggs which are ovulated by the sow. It has been previously assumed that chance is the predominant factor in the final sperm–egg encounter and that sperm motility is the determining factor in success or failure. Although there is a dearth of evidence to substantiate this point, it is now believed that various selective mechanisms exist (Bateman, 1960) to sift out some types of sperm cells in preference to others. In addition it has been suggested that the egg plug of ova and cumulus cells shed by the sow 'attract' and hold the sperm cells until penetration of the zona pellucida has been achieved. Despite this view it still seems a plausible hypothesis that the function of the flagella is important and probably gives sperm the ability to seek out ova in the ampullae thereby increasing the probability of sperm–ova collision.

The first obstacle the sperm cell has to negotiate following a collision with an ovum is the layer of cumulus oophorus cells. If satisfactory capacitation has occurred then the sperm cell digests its way through the cumulus layer by the secretion of hyaluronidase from the loosened acrosome which breaks down the hyaluronic acid matrix of which the cumulus layer is composed.

Upon reaching the zona pellucida of the ovum the sperm may be anchored to the membrane as a result of the secretion of some as yet unidentified substance from the ovum. The sperm then begins its passage through the zona pellucida facilitated both by flagella action and by a proteolytic enzyme secreted from within the acrosome. It is at this stage that the acrosome is lost exposing the inner membrane.

A zona reaction has been observed in the pig preventing the passage of further sperm cells into the ovum but in contrast to other species extra sperm cells may penetrate the zona pellucida but not succeed in swimming through it. It may thus be inferred that the zona reaction is established upon the sperm cell reaching the vitellus and this sets off a biochemical change in the zona membrane and perivitelline space inhibiting the further transport of supplementary sperm.

The final phase in sperm penetration is triggered by the attachment for possibly 30 minutes of the sperm cell to the vitelline membrane. This attachment 'activates' the ovum or initiates the second reduction division of the nuclear material within the ovum. As the sperm cell penetrates the vitellus, its tail is shed (the fate of the mitochondria which the tail region consists of has never been established). Recent scanning electron micrograph studies have indicated that, in penetrating the vitelline membrane, microvilli from the plasma membrane of the ovum grasp the sperm head and, with the simultaneous rupture of sperm plasma membrane and ovum membrane, fuse to encapsule both sperm and egg nuclear material. Once complete this process stimulates the vitelline block or the inhibition of entry of further sperm through the vitellus. Polyspermy or the appearance of supernumerary sperm cells within the vitelline membrane may also occur in about one or two per cent of cases. Where polyspermic fertilization occurs, the chances of the embryo subsequently surviving beyond the early cleavage divisions is remote and certainly implantation is unlikely to occur. Hence polyspermy is a lethal condition lowering reproductive efficiency. The zona reaction and vitelline block have evolved to prevent this condition reaching significant proportions. It has been observed however that aged ova are associated with a reduced zona reaction and vitelline block resulting in a high incidence of polyspermy. This has some bearing on the need to time insemination with precision to achieve maximum effective fertilization rate.

When activation occurs the vitelline membrane shrinks in volume and much of the fluid within the vitellus is projected into the perivitelline space. The sperm head changes its shape and texture and the perforatorium and tail are cast. A number of nucleoli appear and subsequently coalesce. A nuclear membrane appears around the chromosomal material. At the same time a similar female pronucleus is forming and just before syngamy starts the two pronuclei (from sperm and ovum) are of roughly equal volume.

Eventually these two pronuclei come into contact and begin to be absorbed into each other in the fusive process of syngamy. The nuclear membrane disappears and the two sets of chromosomes unite to re-establish the diploidal cell complement. This is the prophase stage of the first cleavage division and fertilization is now complete. In just the same way as any body cell divides and replicates itself, the cell then undergoes and completes the mitotic division and pregnancy is established. The duration of fertilization in the pig from first penetration of the sperm to metaphase of the first cleavage division is around 12–14 hours.

EGG TRANSPORT

At ovulation the fimbriae of the oviduct are engorged with blood and extended. This aids the collection of eggs shed from the ovary. If timing of insemination has been correct the eggs should be fertilized rapidly and subsequently these fertilized ova begin the gradual descent through the oviduct until they reach the uterus some 50 hours or so after ovulation.

The passage of eggs down the ampullae is fairly rapid but at the ampullae–isthmic junction progress is stopped for quite some time before a more gradual rate of passage begins again down the isthmus to the uterotubal junction.

The transportation of eggs is made possible by both the contraction of smooth muscle in the oviductal wall and by the ciliated epithelial layer projecting into the lumen of the oviduct. The rate of passage is controlled by the constrictions which occur at the uterotubal junction and the ampullae–isthmic junction. In turn, the workings of this and other physiological factors are controlled by endocrine status and in particular the balance of progesterone and oestrogen in the circulation. Certain prostaglandins are also known to have an influence on the contractile ability of the oviduct but the precise role of these in egg transport is as yet unknown.

6.2 Factors affecting fertilization rate

With the right conditions, fertilization rate in the pig is high and well over 90 per cent in most estimates (Casida, 1954; Haines, Warwick and Wallace, 1959; Perry and Rowlands, 1962). Hancock (1962) has also pointed out that more often than not fertilization losses are due to whole fertilization failure in a number of animals and in the bulk of sows fertilization rates are 100 per cent. Animals which do exhibit complete fertilization failure will in fact be observed to return to heat 21 days after the original insemination and therefore this type of reproductive loss contributes to conception rate losses seen in practice.

Within the normal range of ovulation rates there is no effect of ovulation rate on fertilization losses (Hunter, 1964) but there is evidence that for an excessive number of eggs shed at ovulation, there is an associated decrease in the proportion of eggs developing normally immediately following fertilization (Hunter, 1966) and the percentage of fertilized eggs drops to around 84 per cent for ovulation rates in the region of 37. Abnormal embryos at this stage fall into a number of categories: fragmented embryos, polyspermy, secondary oocytes unpenetrated and primary oocytes. Gonadotrophin treatment and its associated effect on ovulation rate often causes the release of oocytes from follicles which would not normally ovulate at that time. They may be genuinely immature oocytes or they may in fact have been residing in atretic or potentially atretic follicles and failed to undergo their first maturation division at the normal time (Hunter, 1966).

Wrathall (1971) has put forward a hypothesis to explain the observed level of polyspermy in superovulated animals. He suggests that 'when ovulation is spread over a longer period than normal, progesterone levels may be rising before the last follicles actually ovulate. Thus the last eggs to be ovulated might be subjected to an adverse steroid environment almost as soon as they enter the oviduct and be incapable of preventing polyspermy. The so-called "ageing" process of eggs may therefore be at least partly due to the influence of progesterone'. Progesterone-treated superovulated gilts have in fact shown an improved fertilization rate when progesterone was followed by oestrogen (Dzuik and Polge, 1965) and so it seems that these two steroids must be studied together to obtain a meaningful result.

6.3 Factors affecting conception rate

THE BOAR AND TIMING OF INSEMINATION

Boender (1966) has established that the use of A.I. can be associated with a 10–25 per cent reduction in conception rate (non-return percentage) when compared to natural service. This observation is probably largely as a result of the practical difficulties in carrying out the A.I. technique at the right time relative to ovulation. When a boar is on hand, although there is still a possibility of getting the timing wrong, the probability is better of getting it right.

The conception rate achieved with natural service can be markedly affected by the sire used (Boender, 1966). In other words one boar may be consistently associated with lower conception rates than another (Minkema, 1967; Rahnfield and Swiestra, 1970). With natural service as well as with A.I. the frequency of insemination appears also to be as important as the timing although if a large number of inseminations are made over the course of the heat period then the chance of at least one of these being at the right time must increase considerably. Optimum conception is achieved when insemination is carried out about 12 hours before ovulation (Dziuk, 1970).

It is known also from the work of Du Mesnil du Buisson *et al.* (1970) that ovulation takes place in the second half of oestrus between 38 and 42 hours after the onset of oestrus. The best time for insemination, therefore, will be about 28 hours or a day after the onset of heat. In practice a satisfactory conception is obtained by mating twice on consecutive days at the appearance of oestrus so ensuring 'fresh' semen in the oviducts as near as possible to the theoretical optimum time. Single matings either side of this optimum time will tend to be associated with reduced conception and farrowing rate (Dzuik, 1970).

SEASONAL EFFECTS

Scofield and Penny (1969) found no seasonal effect on conception rates but Nedeleniuc and Dinu (1973) working in Eastern Europe have reported a significant drop of around 10 per cent in conception rate for matings occurring in June, July and August in contrast to matings occuring in November and December.

Components of seasonal effects are of course changing light patterns and temperature and little evidence exists to separate these two effects. It is known, however, that poor fertilization may accrue from high ambient temperature at mating (Sainsbury, 1971) and again the effect probably is an all or nothing effect causing in some sows complete loss of fertilization and return to service 3 weeks later. This effect is difficult to evaluate for the female due to the confounding effect of high ambient temperature on the boar's performance.

Currently the majority of sows are housed in controlled environment conditions often with a lighting regime dependent upon how long the stockman leaves on the lighting system. Hence although significant seasonal effects exist they may be of little importance as a result of this housing practice. As yet the optimum temperature and lighting pattern for maximizing conception rate has not been established but evidence from the National Agricultural Centre pig unit at Stonleigh, England has tentatively suggested that a lighting regime of 12 hours on and 12 hours off in the service house gives best results. No recommendations are available at present giving lighting patterns for dry sow houses.

THE EFFECTS OF FEED INTAKE

Scale of feeding between weaning and remating was shown by Brooks and Cole (1972) to influence conception rate. A feed level of 1.8 kg/day was associated with a drop of 12.5 per cent in conception rate when compared to sows fed 3.6 kg/day between weaning and remating. There are however no further reports to substantiate or refute this observation and in view of the magnitude of the observed effect it is obviously an area worthy of further investigation.

AGE AND GENOTYPE

Little information exists relating conception rate to genotype, age or weight change but it is probable that all of these factors play a significant role in the determination of conception rate. It is known that breeds having a high ovulation rate are associated with a high degree of whole litter loss between fertilization and day 12 of pregnancy (Warnick, Grummer and Casida, 1949). In addition, we know that ovulation rate and parity are closely linked and therefore there will almost certainly be an indirect effect of age (and/or parity) on conception rate. It would appear that any influencing factor which causes an increased ovulation rate has a concomitant effect in reducing conception rate. This has been noted in gilts by Hafez (1952) who has reported that better nutrition improves ovulation rate but can lead to poorer conception rates.

Body condition and weight change are often quoted as being important determinants of ability to hold to service but in the authors' experience the sow can suffer very wide variations in weight change pattern with little adverse effect on ability to conceive (Varley and Cole, 1978). However, there is probably a threshold level of body condition at the top and bottom end of the range outside of which the sow or gilt will have a high propensity to exhibit reproductive malfunction and may repeatedly fail to hold to service.

ARTIFICIAL INSEMINATION

It is certain that the use of artificial insemination predisposes sows to show a reduced conception rate. It is not certain whether or not this effect is inherent in the technique of artificial insemination and what is the magnitude of the reduction in conception rate. Meat and Livestock Commission survey work in United Kingdom herds using A.I. has demonstrated that the response is extremely variable. Some A.I. users do not experience any reduction at all in conception rate whilst others experience a drop of 10–15 per cent or even 20 per cent. This will be dealt with more fully in Chapter 14 but this observation comes down to the fact that some producers have the competence to achieve correct timing of insemination and good technique. Others have difficulty in getting it right and poor conception is the result.

LACTATION LENGTH

Lactation length is also a factor which significantly affects conception rate. This is dealt with in Chapter 9.

6.4 Conclusions

There is a considerable degree of variation between herds for conception rate. In well managed herds 90–95 per cent plus is being achieved and in poor herds it may be as low as 50 per cent. Even within herds there is probably a wide variation from one time to another. Much of the variation can be ascribed to managerial factors rather than indigenous biological limitations imposed by the sow's reproductive physiology. There seems no logical reason why producers should not show higher average conception and farrowing rates and adherence to the normal rules of good husbandry at and around mating should yield better results. Wrathall (1971) has indicated that the following proportions of sows can be apportioned to the various causes of conception failure:

(1) About 5–10 per cent of litters may be lost as a result of fertilization failure (probably more in gilts than sows).
(2) About 5 per cent of litters may be lost as a result of complete embryonic death before or shortly after day 12 of pregnancy (probably more in sows than gilts).
(3) Up to 5 per cent of litters may be lost during the remainder of gestation.

These losses may be due to the factors reviewed in the present chapter but at the farm there is not much a producer can do about them within the constraints of the genotype of his sows and the age distribution of the herd. Feeding and weight change is under direct management control and the old adage of 'fit not fat' is probably a workable rule of thumb to feed by. With larger herds, however, and in view of the fact that the use of automatic individual feed dispensers in sow houses is currently proliferating, the need for a more precise feeding system is self-evident. Hillier (1978) has suggested that there is a close link between weight change (and hence body condition) and reproductive performance. The weighing of sows at regular intervals should help to maintain a regular and optimum growth pattern for *all* sows and not just a proportion of them. Obviously no single feeding scale pattern is suited to all conditions but the general aim might be a weight gain between subsequent parities for sows of about 12–15 kg.

Timing of insemination is by far the most important factor in successful fertilization and conception. Tied in with this is the need to know exactly when a sow shows first oestrous symptoms. A regular, efficient and frequent heat detection programme therefore would seem to be a necessity for good performance to be achieved. When a sow or gilt first exhibits oestrus, a delay of maybe 8 hours ideally should be made before the first insemination and then a second 'insurance' insemination should be made 12–18 hours later to obtain best results. The difficulty in practice is that stockmen tend to detect heat only once a day and so the first observed oestrous symptoms may in fact be at a stage well into oestrus. Insemination is therefore carried out as soon as heat is observed (or when the semen arrives in the case of A.I.) and may certainly be the wrong time for a large proportion of sows.

Without doubt the careful supervision of service is of paramount importance to ensure satisfactory fertilization and conception. To facilitate this a well-designed service area is needed. Currently purpose built service houses are a new concept and no specific design has emerged as better than others. Design criteria should include the logistic aspects of moving sows in and out of boar pens quickly and easily and also housing in close proximity (sight, sound and smell) of boars.

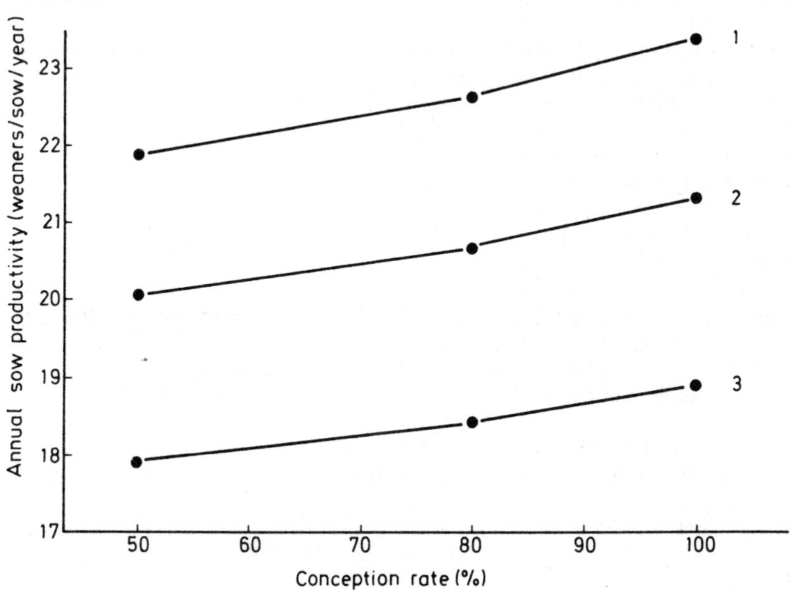

Figure 6.1 The effect of conception rate on annual sow productivity. Curve 1, 3-week weaning; curve 2, 5-week weaning; curve 3, 8-week weaning

Figure 6.1 shows the likely effect of different conception rates on annual sow output for three different weaning ages. Clearly to achieve a conception rate of 90 per cent and over must be a base point for the operation of a high output breeding herd.

6.5 References

AUMAITRE, A. (1972). *23rd Annual Meeting E.A.A.P. Commission in Pig Production (Verona)*

AUSTIN, C.R. (1969). In *Advances in the Biosciences 4*, pp. 5–11. Vieweg, Berlin, Pergamon Press

BATEMAN, N. (1960). *Genet. Res., Camb.* **1**, 226–238

BOENDER, J. (1966). *Wld Rev. Anim. Prod.* **2**, 29

te BRAKE, J.H.A. (1972). *23rd Annual Meeting E.A.A.P. Commission in Pig Production (Verona)*

BROOKS, P.H. and COLE, D.J.A. (1972). *Anim. Prod.* **14**, 241

CASIDA, L.E. (1953). In *Pregnancy Wastage* (Ed. by E.T. Engle-Springfield). Illinois, Charles C. Thomas

DU MESNIL DU BUISSON, F., MAULEON, P., LOCATELLI, A. and MARIANA, J.C. (1970). *Colloq. Ste. Nle. étude Steril. Fertil.* p.225. Paris

DZUIK, P.J. (1970). *J. Reprod. Fert.* **22**, 277–282

DZUIK, P.J. and POLGE, C. (1965). *Vet. Rec.* **77**, 236–238

HAFEZ, E.S.E. (1952). *J. agric. Sci., Camb.* **54**, 170

HAFEZ, E.S.E. (1973). In *The Biology of the Cervix* (Ed. by R.J. Blandeau and K.S. Moghissi), pp. 23–56. Chicago, University of Chicago Press

HAFEZ, E.S.E. (1976). *Reproduction in Farm Animals*. Philadelphia, Lea and Febiger

HAINES, C.E., WARWICK, A.C. and WALLACE, H.D. (1959). *J. Anim. Sci.* **18**, 347–354

HANCOCK, J.L. (1958). In *Studies on Fertility* (Ed. by R.G. Harrison). Vol. 9, pp. 146–158. Oxford, Blackwells Scientific

HANCOCK, J.L. (1962). *Anim. Breed. Abstr.* **30**, 285–310

HARING, F. (1937). *Züchtungskunde* **12**, 1

HILLIER, M. (1978). *Pig Fmg* **26(1)**, 68

HUNTER, R.H.F. (1964). *Anim. Prod.* **6**, 189–194

HUNTER, R.H.F. (1966). *Anim. Prod.* **8**, 457–465

HUNTER, R.H.F. and LEGLISE, P.C. (1971). *J. Reprod. Fert.* **24**, 233–246

MINKEMA, D. (1967). *Veeteelt-en Zvivelber* **10**, 161–172

MOGHISSI, K.S. (1972). *Fert. Steril.* **23**, 295–306

NEDELENIUC, V. and DINU, I. (1973). *Lucravile Stunfifice ale Statiunu centrale de cercetari pentru cresterea porcilor peris.* **1**, 147–151

PERRY, J.S. and ROWLANDS, I.W. (1962). *J. Reprod. Fert.* **4**, 175–188

RAHNFIELD, G.W. and SWIESTRA, E.E. (1970). *Can. J. Anim. Sci.* **50**, 671–675

ROWLANDS, S.W. (1957). *J. Endocr.* **16**, 98–106

SAINSBURY, D.W.B. (1971). In *Pig Production* (Ed. by D.J.A. Cole). London, Butterworths

SCOFIELD, A.M. and PENNY, R.H.C. (1969). *Br. Vet. J.* **125**, 36–45

SELF, H.L., GRUMMER, R.H. and CASIDA, L.E. (1955). *J. Anim. Sci.* **14**, 573–592

SQUIERS, C.D., DICKERSON, B.E. and MAYER, D.T. (1951). *Res. Bull. Mo. agric. Exp. Stn* **494**

VAN DER HEYDE, H. (1972). *Proc. Br. Soc. Anim. Prod.* **1**, 33–36

VARLEY, M.A. and COLE, D.J.A. (1978). *Proc. Br. Soc. Anim. Prod.* 1978 Winter Meeting, Harrogate

WARNICK, A.C., GRUMMER, R.H. and CASIDA, L.E. (1949). *J. Anim. Sci.* **8**, 569–577

WRATHALL, A.E. (1971). *Prenatal survival in Pigs*. Slough, England, Commonwealth Agriculture Bureau

Chapter 7

Pregnancy

Approximately two-thirds of a sow's life is spent in gestation and so the resources devoted to the sow throughout pregnancy are an important consideration and determinant of the success or failure of a system. With modern housing for pregnant sows based on stall or tether systems over slatted floors the labour input throughout pregnancy is at a low level and the danger is that it falls below a threshold level after which the system begins to deteriorate in terms of sow health status, body condition, heat detection and pregnancy maintenance. Particularly on the larger unit individual sow observation and attention is still a must in the dry sow house to maintain good results. In addition to labour input, feed input in pregnancy represents the major feed cost to the breeding herd and as such is critical to the optimum economic performance of the herd.

Pregnancy or gestation begins at fertilization and as we saw in the previous chapter the fertilized ovum passes down the oviduct, cleaving as it goes, and appears in the uterus at about day 4 post coitum at the morula stage.

In the ensuing 3 weeks the ova change from 'self-supporting' eggs into rapidly developing embryos implanted on the uterine wall and becoming increasingly dependent on maternal blood supply. This process of attachment or implantation begins at day 12 or 13 post coitum and is complete by about week 4 of gestation. After implantation the embryos pass on into the fetal stages and finally pregnancy is terminated by parturition. Throughout this whole period losses of embryos and fetuses are due to a multiplicity of diverse factors which will be reviewed towards the end of the present chapter. The extent of this loss is probably the major limiting factor to the eventual litter size of the sow and to a certain extent the gilt. As yet the state of knowledge in this area although extensive is by no means complete.

Generally the length of the gestation period is a non-variable factor and is unaffected by any external stimulus or the size of the litter carried as it is in other polytocous species. It has a mean value for the British White breeds of about 114 days (Braude, Clarke and Mitchell, 1954) although the range can be from 110 to 120 days. Burger (1952) has reported no difference between the Large White breed (113.35 ± 1.42 days) and the Large Black breed (113.66 ± 1.42 days). Fairly recently we have gained a more precise understanding of mechanisms controlling parturition and it is now possible to induce farrowing on a set date by the use of exogenous prostaglandin. This aspect of the sow's reproduction will be covered at the end of this chapter.

7.1 Physiological mechanisms in gestation

ENDOCRINE AND OVARIAN CHANGES

From conception the gonadotrophic hormones in the circulating plasma drop to basal levels by the first day post coitum. Thereafter throughout the whole gestation both FSH and LH appear to stay at very low levels and no marked changes in the pituitary gland concentration have been observed apart from a slight drop in LH concentration in the latter part of pregnancy (Day *et al.*, 1959; Parlow, Anderson and Melampy, 1964; Melampy *et al.*, 1966). This indicates that there is no secretion of gonadotrophins into the blood stream throughout pregnancy and in addition no gonadotrophins are being produced and stored in the pituitary until pregnancy has been terminated.

Progesterone plays the major hormonal role throughout gestation and peripheral levels build up very rapidly in the first 10 days following fertilization (Short, 1960; Rombauts, Fevre and Tarqui, 1971; Edgerton and Erb, 1971; Shearer *et al.*, 1972). Peak values are observed at around 10 days post coitum but by day 20 progesterone levels fall off slightly and remain fairly constant for the rest of gestation until a significant drop occurs shortly before parturition (Shearer *et al.*, 1972). Unlike some other species the pig's only source of progesterone is the ovary and the maintenance of pregnancy depends entirely on luteal function. Ovariectomy at any time during pregnancy results in abortion (Du Mesnil du Buisson and Danzier, 1957). Ovarian secretion of progesterone is in turn dependent on pituitary support and hypophysectomy during pregnancy results also in luteal regression and abortion (Du Mesnil du Buisson, 1966).

Oestrogen secretion in early pregnancy remains at the low level observed after ovulation and conception (Raeside, 1963a; Rombauts, Fevre and Tarqui, 1971) but a significant peak is seen about one month into gestation (Lunaas, 1962) which is short-lived and thereafter levels drop very quickly again.

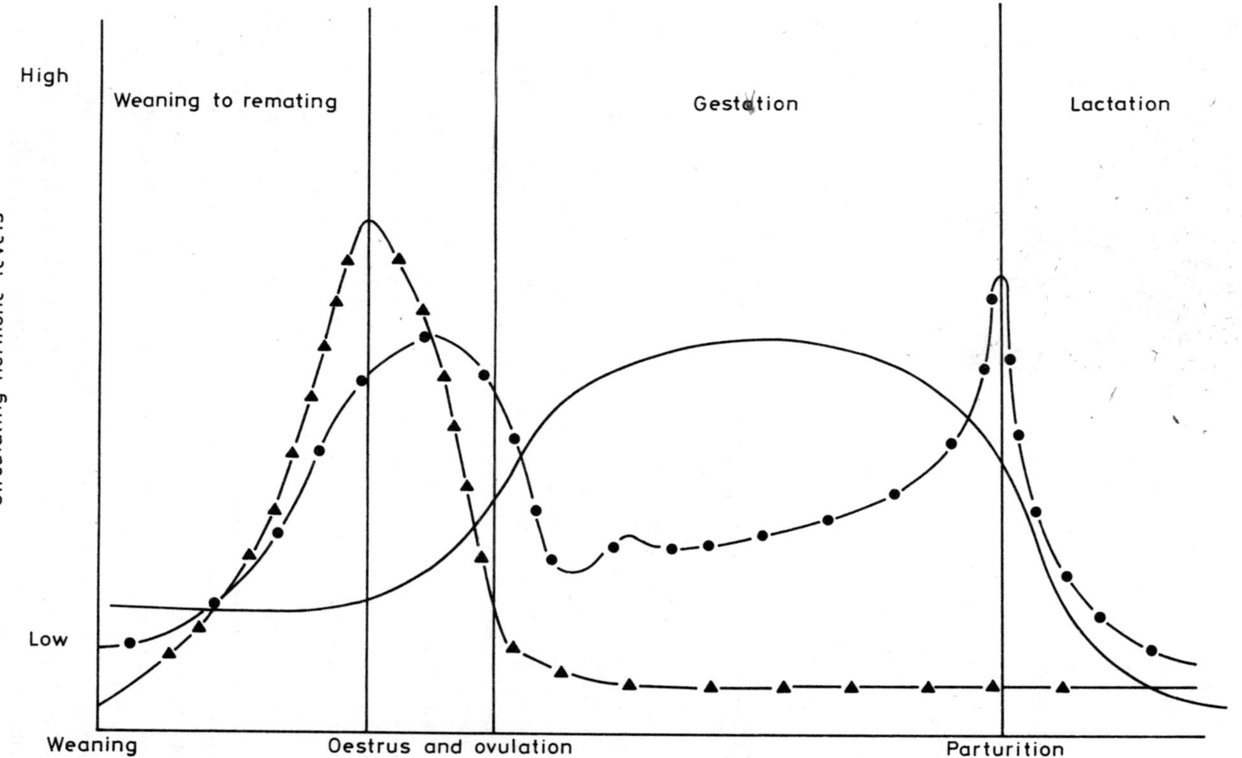

Figure 7.1 Circulatory hormone levels during the reproductive cycle of the sow. Solid curve, progesterone; ● oestrogen; ▲ FSH, LH

From about the tenth week of gestation plasma oestrogen levels begin to rise and in the last two weeks of pregnancy there is a massive secretion of oestrogen (Raeside, 1963b; Rombauts, Fevre and Tarqui, 1971). No change in ovarian morphology has been reported during this time (Melampy *et al.*, 1966) and it seems likely that the oestrogen is feto-plancental in origin. The exact function of this large increase in oestrogen secretion at or just before partutition is not clear but the effects on the reproductive tract and the behaviour of the sow immediately after parturition are well known (Burger, 1952; Warnick, Grummer and Casida, 1949).

The uterus responds rapidly to these steroid balance changes and the endometrium becomes quite oedematous in response to high levels of circulating oestrogens just prior to parturition (Corner, 1921).

These hormonal changes are represented schematically in *Figure 7.1.*

7.2 Changes in the reproductive tract during gestation

Although not quite as apparent as in other species the vagina and vulva become more oedematous as pregnancy progresses and the level of vascularization increases, particularly in the later stages of gestation.

A very viscid mucus is secreted by the cervix and the so-called mucous plug of pregnancy is established within the cervix serving in part to protect the developing fetuses from external action. Just prior to parturition this mucous plug breaks down and is discharged via the vagina.

As pregnancy develops the uterus gradually expands. The fetuses and uterine contents become more voluminous and as this occurs the muscular layers of the uterine wall remain inactive to prevent premature expulsion of the litter.

The expansion of the uterus is in three phases. The first phase is characterized by endometrial proliferation which occurs before the attachment of blastocysts. After implantation the second phase commences when growth by muscular hypertrophy and increase of connective tissue is seen. The last phase is a 'stretching' phase where cells already present are distended by the rapidly growing contents within the uterus.

Changes in the ovary during pregnancy are centred on the establishment of the corpora lutea just after ovulation and the maintenance of luteal tissue in gestation by a luteotrophic stimulus probably originating from the embryos.

During a non-fertile oestrous cycle a luteolytic signal from the uterus (prostaglandin) causes the corpora lutea to regress (*see* Chapter 5). If viable embryos are present in the uterine horns, however, the luteal tissue overcomes the uterine luteolytic influence and any further oestrous cycling is inhibited. It would appear that there must be four or more embryos present within the uterus of the pig or this effect is not invoked (Wrathall, 1971) and embryos degenerate and pregnancy terminates.

If, for some reason, a large portion of the uterus is not occupied by any embryos again pregnancy ceases. If the empty portion is removed surgically, however, pregnancy is maintained despite the regression of the corpora lutea on the ovary adjacent to the surgically removed portion. This strongly suggests that just before 12 days post coitum the embryo emits some physiological signal which induces the maintenance of the corpora lutea of pregnancy.

Towards the end of pregnancy under the influence of rising oestrogenic hormones and relaxin, relaxation of the pelvic ligaments occurs. In gilts the pubic symphysis undergoes demineralization and some separation is seen at parturition.

One of the overt signs of pregnancy is the live weight change of the gilt and sow as a result of increase in uterus and uterine contents and also as a result of real gain in body weight. This effect obviously is largely under the control of feed intake. Certainly the phenomenon of pregnancy anabolism is very evident in the female pig and given a generous feed allowance she will increase her live weight significantly as a consequence of steroidal balance and enhanced feed efficiency.

7.3 Embryology of the pig

The embryology of the pig from cleavage through to the final fetal stages is a highly complex process and it is not intended in this section to cover all the descriptive and analytical knowledge because this has been covered elsewhere (Patten, 1953). It is however worth looking at this point at some of the more important features of embryonic development.

CLEAVAGE

The fertilized ovum rapidly cleaves after syngamy using its own energy resources as it passes down the oviduct and into the uterine horn. At the commencement of this stage the free living embryo is a single cell of comparatively large volume and with a high ratio of cytoplasmic to nuclear material. At the end of the cleavage stages, although there has been no growth in terms of increase in the dimensions of the whole embryo, the nuclear to cytoplasmic ratio returns to normal somatic cell levels as the numbers of cells are multiplied.

There is no apparent systematic plane to the first division but the second division occurs at right angles to the first and so on. The whole process is not in exact synchrony so we may observe, at the same point in time, embryos at different stages in the same reproductive tract.

At the 16–32 cell stage, the cells are bunched together in a solid ball of cells surrounded by the zona pellucida and is referred to as a morula.

Much work recently has examined the fate of daughter cells in cleavage and the current belief is that in mammals cleavage is of the indeterminate type. That is, the fate of any one particular cell in early cleavage cannot be designated as the precursor of say lungs or skin but depends on the chance spatial position of a cell within the embryo. It has been shown in the laboratory species that any cell within a cleaving embryo has the capacity if cultured correctly to be the progenitor of a whole new embryo.

In the pig, the inner cell mass appears to develop not from the inside cells of the morula but from a group of larger less actively dividing cells at one pole of the embryo. *Table 7.1* gives the approximate phasing of the cleavage stages in the pig.

Table 7.1. Stages in cleavage

Time relative to ovulation	*Cleavage stage*
+ 14–16 hours	2 cell
+ 2 days	8 cell
+ 2 days	Into uterus
+ 5 days	Late morula

By the late morula stage the embryo consists of many hundreds of apparently undifferentiated cells and by this stage the embryonic genome is functioning actively to control its own rate of development and protein synthesis.

BLASTULATION, SPACING AND EMBRYO METABOLISM

Blastulation is heralded by the appearance within the solid ball of morula cells of a blastocoele or fluid-filled cavity and differentiation of cells into trophoblast and inner cell mass. At the same time (between 8 and 11 days post coitum) the embryos undergo a migratory activity both within and between the uterine horns (Dzuik, Polge and Rowson, 1964; Dhindsa, Dzuik and Norton, 1967; Polge and Dzuik, 1970). The mechanism controlling embryonic migration and spacing has never been elucidated and although it was once thought that spacing played a role in optimum survival rates for embryos, this belief has now been refuted and overcrowding of part of the uterus up to about day 35 of pregnancy is not associated with excessive embryo loss (Webel and Dzuik, 1970).

Following and during spacing of embryos the trophoblast wall undergoes rapid elongation along the length of the uterine horns as implantation approaches.

At blastulation the embryo is in a precarious position as far as its energy metabolism is concerned. It may well be that a proportion of losses of embryos occur at this critical time as a result of embryos being unable to develop normally due to insufficient nutrition, which is in part controlled from within the embryo and in part controlled by its external environment (the uterus). Much of the work on the mechanisms of early development has been caried out on the laboratory species but as Brinster (1974) has pointed out, all mammalian embryonic functions follow a strikingly similar course particularly in the very early period and extrapolation to the pig is therefore valid.

At blastulation an important characteristic of the embryo appears which is the ability to accumulate fluid and this ability depends on two cellular functions. One is the transport of various substances and the second is selective permeability. Between about day 4 and day 8 the embryo increases in volume 4000-fold and since the dry weight of the blastocyst is only 1 per cent of net weight (Lutwak-Mann, 1959), this increase must be due to uptake of water. It has also been observed (Cross and Brinster, 1969) that there is an electric potential across the trophoblast wall between days 5 and 7 post coitum. This is good evidence to suggest that active transport of ions across the trophoblast wall from uterine fluids exists. Both chloride and sodium ions have now been identified and quantified as being readily transported in equal proportions and bicarbonate ion has tentatively been suggested as being taken up in significant proportions (Brinster, 1974). Bicarbonate ion is also present in high concentration in oviductal, uterine and blastocoele fluids and is probably very important in early embryonic metabolism (Brinster, 1973). Energy metabolism is in fact based on glycogen in the early cleavage stages (Stern and Biggers, 1968; Thompson and Brinster, 1966). Biochemical studies have indicated that the embryo synthesizes all its own glycogen by the eight cell stage and then uses a small part of this glycogen for blastocyst formation (Brinster, 1974). *In vitro* studies of embryos have failed to identify whether the precise source of the carbon is from pyruvate or glucose which is synthesized to glycogen up to blastocyst formation. This problem is as yet unsolved. Glycogen content however can range from 5 per cent to 45 per cent of blastocyst dry weight and depends largely on availability of glucose in the culture media. Conditions in the reproductive tract therefore must have far-reaching effects on embryonic energy mechanisms and hence play a major influencing role in the development and survival of embryos up to and including implantation.

Brinster (1965a) has also shown that optimum development of the embryo is influenced by the pH of the culture media through its effect on concentration of the substrate components (pyruvate, lactate, oxaloacetate and phosphoenolpyruvate). Again this is another pathway by which the embryo may respond to uterine conditions.

Uptake of oxygen by the embryo in the first days of development is largely for pyruvate oxidation in the cleavage stages, but by implantation the energy metabolism has changed from this unusual one to one more reliant on glucose oxidation (Brinster 1967, 1968). Lactate formation as expected also follows glucose oxidation quite closely.

On the basis of relationships observed between pyruvate and lactate in the stages up to implantation it has been postulated (Brinster, 1965b) that it is the ratio of NAD to NADH which regulates embryonic development.

All of these biochemical observations strongly suggest early embryos and blastocysts are in an unstable metabolic position and highlight the importance of optimum uterine conditions if optimum embryo development and survival rates are to be attained. As yet the quantification of 'optimum uterine conditions' has not been established.

Protein synthesis in the fertilized ovum and blastocyst is also in an activated state. Despite the fact that the number of cells is doubling approximately every 12 hours, it does not appear that the new nuclei are functioning to double the amount of message translated into new protein. This suggests that maternal messages are the dominant factor in protein synthesis before the eight cell stage, but after this stage and up to implantation the embryonic genome begins to actively function to control its own protein manufacture.

It would appear therefore that embryonic development is characterized by a spurt of general metabolic activity just before or at the time of blastocyst formation. This stage is crucial if normal development is to continue and disturbances here could result in blastocyst losses and may be responsible for the induction of losses at the later stage of implantation.

7.4 Implantation

Attachment of the embryo to the uterine wall begins at about day 12–13 post coitum and is complete by day 24 of gestation (Crombie, 1970). As mentioned previously, a mechanism exists to evenly space blastocysts prior to implantation. As yet its precise nature is not known although uterine contractions are probably involved. Certainly when one blastocyst implants close to another this appears to have no detrimental effect on the immediate development of either but in the fetal stages maternal blood supply may limit the development and survival of both. There is some evidence also to suggest that the stimulus of an embryo implanting causes a proliferation of cells and growth of the endometrial cells in the immediate vicinity of the implanting embryo thus causing a separation of implanting embryos as gestation progresses.

The orientation of pig embryos at implantation is such that the embryonic disc is always situated on the antimesometrial side of the uterine horn. By the end of pregnancy however fetuses may be facing one way or the other up the horn.

The uterine wall is pre-sensitized for implantation by the balance of steroid hormones. Oestrogen followed by rising progesterone leads to the endometrial lining being at the optimum state to accept blastocyst attachment. Any disturbance in endocrine balance therefore inhibits implantation rate.

After the initial contact between embryo and endometrium at day 12 post coitum the trophoblast begins to proliferate very rapidly and begins to invade the endometrium. The trophoblast wall folds as a result of the rapid expansion of the trophoblast not being matched by uptake of fluid into the blastocoele. The endoderm appears with the onset of gastrulation and very rapidly the whole blastocyst elongates to perhaps 50 cm with the embryonic disc occupying a short but enlarged section of the tube in the centre of each blastocyst. By day 24, mass invasion by the trophoblast of the endometrial lining of the uterus has taken place as the gradual establishment of the diffuse epitheliochorial placenta begins. This is almost the same manner of growth as that of tumours invading other body tissues. There is evidence to suggest the uterus contains an inhibitory substance at this time to control and regulate this growth and this unknown inhibitory substance is removed under the influence of oestrogen (McLaren, 1973). It has been well demonstrated that away from this inhibiting factor (i.e. ectopic sites or *in vitro*) trophoblast growth occurs more easily even in the presence of hormonal influences.

Throughout implantation and up to the onset of placental function the embryo depends for its nutrition on uptake of 'uterine milk' or histotroph. One might presume that as the blastocyst is undergoing such rapid growth and development at this stage it is highly sensitive to uterine conditions. If optimum conditions do not prevail then successful implantation may not occur. Of all losses in the embryonic stage, the bulk are in fact lost at or around implantation and may be the result of inadequate embryonic nutrition. Although there have been some recent attempts to influence uterine milk content by exogenous means, no effect has been observed on the survival of embryos. This may in the future prove a promising technique for enhancing reproductive output in the pig.

By the time implantation is complete the uterine wall is deeply folded and the chorion is in close contact with the uterine epithelium to facilitate capillary exchange.

7.5 Gastrulation and organogenesis

Gastrulation involves the cells of the embryonic disc. Relative movement of groups of cells is seen and eventually these are aligned into the necessary formation in readiness for the ensuing development of the various body tissues and organs. This process converts the embryo into a three-layered structure from which the *fetus* develops. Cells migrate and split off from the inner cell mass to give rise to the notochord, endoderm, mesoderm and ectoderm.

Table 7.2. Important phases of the embryonic development in the pig

Stage	*Time after conception* (days)
Morula	3.5
Blastula	4–5
Gastrulation	7–8
Elongation of chorionic vesicle	9
Primitive streak formation	9–12
Open neural tube	13
Somite differentiation (first)	14 (3–4 somites)
Fusion of chorio-amnionic folds	16
Heart beat apparent	16
Closed neural tube	16
Allantois prominent	16–17
Forelimb bud visible	17–18
Hindlimb bud visible	17–19
Differentiation of digits	28+
Nostril and eyes differentiated	21–28
Implantation	12–24
Allantois replaces all of exocoelom	25–28
Eyelids close	28
Hair follicles apparent	28
Tooth eruption	16 mm (Fetal length)
Birth	112–118

From Hafez (1976).

From this three-layered embryo is formed the neural tube and this is the progenitor of the brain and spinal chord. Following this stage there is a gradual and systematic formation of various body organs and structures. The chronological occurrence of the important stages in embryonic and fetal development are presented in *Table 7.2.*

7.6 Placentation

From implantation the placenta becomes established to facilitate physiological exchange of materials to sustain the fetus. In the pig this physiological exchange occurs over the whole connecting surface of maternal (endometrial) and fetal (chorionic) surfaces and villi project from both surfaces to increase the surface area in contact, hence aiding the passage of materials. This type of placentation is known as diffuse epitheliochorial placentation unlike the sheep and cow which have specialized placentomas for the purpose of exchange.

Electron microscopic studies have revealed that the junction between fetal and maternal tissues in the placenta of the pig is the interdigitation of microvilli from the trophoblast and uterine epithelium.

The formation of chorionic villi is an important step in the development of the placenta. These consist of vascular mesenchymal cones surrounded by cuboidal trophoblastic and giant binucleate cells.

The process of haemotrophic nutrition (the feeding of the developing fetuses via the maternal and placental blood system) is heralded by marked vascularization of all the structures involved in placentation. Not all the blood entering the placenta participates in gaseous exchange, however, due to the presence of shunts in maternal and fetal umbilical circulation. The blood of fetuses and dam never come into direct contact yet the two systems of circulation are in such close proximity that oxygen and nutrients can pass from the maternal blood to the fetal system and waste products pass in the opposite direction. Simple diffusion or the movement of molecules from an area of high concentration to an area of low concentration was once thought to be the more important mechanism in placental transfer. It is now known that most molecules are transferred by active transport across the various membranes and can in fact be 'pumped' against a concentration gradient. A good account of these physiological processes is given by Metcalfe, Bartels and Moll (1967).

7.7 Parturition

Parturition is defined as the process of the pregnant uterus delivering the fetuses and placenta from the maternal organism. The process is under direct hormonal control with both fetal and maternal endocine systems taking part and the whole process can be considered in distinct stages.

ENDOCRINE CONTROL OF PARTURITION

It was once thought that the process of parturition was triggered by the maternal uterine tissues and probably endometrial distension was a factor involved. It is now believed that the initial stimulus to 'set the wheels in motion' comes from within the fetus itself and probably the fetal pituitary gland plays a major role. This view is supported by experiments carried out on sheep following observations made by sheep farmers in the Idaho mountains. At certain times of the year ewes were grazing on alkaloid weed which had the effect of 'knocking out' the fetal pituitary gland. Subsequently many ewes died whilst carrying lambs of very much above average birthweights with no signs of parturition about to commence.

The chain of events leading up to delivery probably originates in the fetal hypothalamus which stimulates the fetal pituitary to release ACTH (adrenocorticotrophic hormone). This promotes corticosteroid production from the fetal adrenal glands. In turn fetal corticosteroids have an effect on the placenta and/or the uterus and stimulate the production of prostaglandin. As yet the precise mechanism whereby glucocorticoids control prostaglandin production has not been elucidated, but certainly the effects of prostaglandin are now well known. Luteolysis rapidly occurs and without luteal support and high levels of circulating progesterone, uterine contractions are no longer inhibited and the myometrium commences rhythmic contractions leading up to labour.

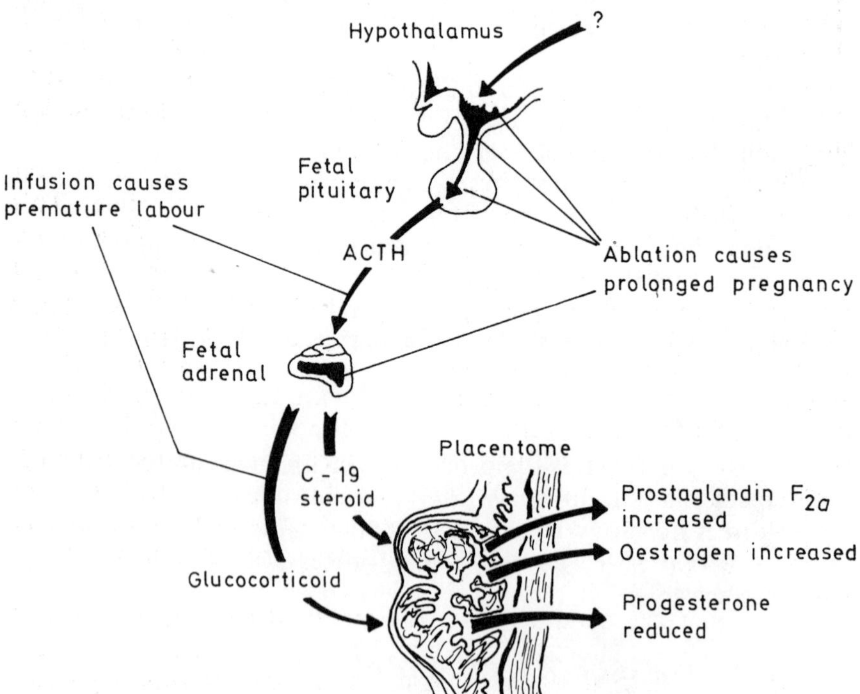

Figure 7.2 The stimuli controlling the onset of parturition (from Liggins *et al.*, 1972)

A schematic diagram is given in *Figure 7.2* to illustrate this sequence of events. It is also important to note here that most of the work in this area has been carried out on the ewe. In all probability the same mechanism applies to the pig.

STAGES OF LABOUR

Parturition can be divided into three stages: (a) preparatory, (b) expulsion of the fetuses and (c) expulsion of the placenta.

The preparatory phase is characterized by a fall in plasma progesterone which begins about 2 days prior to parturition. At the same time an associated rapid rise in plasma oestrogens occurs which is probably placental in origin. As labour ensues, oxytocin from the posterior pituitary gland is released together with the hormone relaxin which helps to dilate the cervix and ease the passage of the conceptus.

A transitional period of irregular and brief contractions is gradually transformed into coordinated and periodic contractions as labour gets under way. In the pig, contractions begin just cranial to the fetus nearest the cervix and the remainder of the uterus stays quiescent. Contractions result from autonomic neural reflex activity on the smooth muscle in the uterine wall and the effect is mediated by oxytocin. At the end of the preparatory stage the cervix has expanded allowing the uterus and vagina to become a continuous tract. Allantoic fluid flows from the vulva following the rupture of the chorio-allantois.

The second stage of labour is initiated by the distended amnion of the nearest cervical fetus being forced into the pelvic inlet. This initiates reflex and voluntary contractions of the diaphragm and abdominal muscles. The fetus is then propelled fairly rapidly through the cervix as the water bags rupture and then out through the vagina to the external environment. The diffused placenta detaches itself *en masse* from the uterus just at the end of the first stage of labour and hence it is imperative that subsequent stages are fairly rapid otherwise many fetuses may die due to hypoxia. Around 60–70 per cent of all fetuses are delivered with anterior presentation although posterior presentation does not appear to be associated with a higher probability of perinatal death. This second stage of labour normally takes between 1 and 4 hours. Outside of this range perinatal mortality rises exponentially.

The third stage of labour sees the placental membranes expelled via the vagina still under the influence of uterine contractions. Where this process is delayed or impeded to any extent metritis may occur and the sow or gilt may suffer postfarrowing fever with high temperature and loss of appetite. In some cases the syndrome M.M.A.

Table 7.3. Timing of parturition

		Preparatory stage	*Expulsion of fetuses*	*Expulsion of placenta*
Time	Range	2–12	1–4	1–4
(hours)	Abnormal	6–12	6–12	24

(mastitis, metritis and agalactia) may also be contracted with obvious detrimental effects on the litter of piglets.

Table 7.3 gives an indication of the time sequences for the various stages of labour in a 'normal' parturition. The behaviour of the sow at and around parturition is very characteristic. Immediately prior to labour she is in a very restless state and under the influence of prolactin and oxytocin, milk secretion and release begins some 12 hours pre partum (English, 1968). Group-housed sows will retire to some quiet corner of the pen and ward off any inquisitive penmates. Sows on straw systems will begin carrying straw around the pen in their mouths and 'nesting' before finally settling down in a prone position before labour commences. While in labour and delivering piglets she shows apparent disregard for piglets born first and, particularly in gilts, cannibalism of piglets may occur. As soon as labour is complete, however, the maternal instincts come to the fore and the sow suckles the whole litter. Certainly some sows do show behavioural problems during labour and high perinatal mortality may accrue without a suitable farrowing crate being used to prevent piglets suffering overlying by the sow.

INDUCTION OF PARTURITION

It has now been demonstrated (Diehl, 1974) that it is technically feasible to induce a sow or gilt to farrow on a pre-determined day and even within a specified time range during the day. This uses the fact that prostaglandin plays a vital role in the normal initiation of parturition through its effect on the corpora lutea of pregnancy. Hence, if infusions of prostaglandin or one of its analogues are given intravenously on day 111 or 112 of gestation then parturition commences between 24 and 26 hours later. This has of course tremendous practical application since due to the circadian rhythm of birth 60–67 per cent of sows will normally farrow at night when there is no stockman in attendance. Consequently high perinatal losses of piglets result. By ensuring that all sows farrow in working hours much of this loss can be avoided. In addition, batch farrowing systems work considerably better by knowing exactly when a group of sows will produce their young.

At present the technique has not been commercially exploited or evaluated on a wide scale but in the near future it seems probable that parturition induction will become an essential feature of pig management.

7.8 Factors affecting embryonic and fetal mortality

Following ovulation, the potential litter size represented by the number of eggs shed gradually diminishes with the cessation of development of a proportion of embryos at different stages throughout gestation. Much of this loss has occurred by the end of week 3 of gestation in the embryonic stage and indeed many estimates of embryo survival are based on observations at day 25 of pregnancy.

Losses at the various stages are due to different reasons and tend to be associated with specific physiological events such as implantation.

EXTENT AND TIMING OF EMBRYONIC DEATH

Until recently the 30–40 per cent of embryo losses occurring have been thought to take place around the time of implantation but some doubt as to the timing of implantation existed. Crombie (1970) has demonstrated that the first signs of attachment of the embryo to the uterus appear on day 13 of pregnancy and implantation is complete by day 24 post coitum.

Perry and Rowlands (1962) reported that 22 per cent of embryos recovered during days 6–9 post coitum were abnormal and yet no abnormal embryos were recovered in the tubal stages. The stage of 6–9 days post coitum is just after the blastocyst has developed and is about to elongate. In the same study, the embryo loss at days 13–18 post coitum was 28.4 per cent and at days 26–40 was 34.8 per cent. Phillippo (1967) in addition found no significant difference between embryo loss at day 13 (41 per cent) and embryo loss between days 15 and 25 (29.9 per cent). Scofield (1969) and Scofield, Clegg and Lamming (1974) studied the period 9–13 days post coitum and concluded that this is a critical phase of development when a large proportion of loss occurs.

It seems therefore that there are two important stages when embryo losses are concentrated: at the stage when the blastocyst begins rapid elongation and also at and around implantation.

LOSSES RELATED TO OVULATION RATE

It has been shown in a number of studies that the level of embryo mortality is dependent on ovulation rate (Perry, 1954; Rathnasbapathy, Lasley and Mayer, 1956; King and Young, 1957) but the

Table 7.4. Relationship between ovulation rate and embryo survival

Author	*Treatment*	*Sows or gilts*	*Ovulation rate*	*Number of animals*	*Post coitum stage of assessment* (days)	*Embryo survival* (%)	*Number of surviving embryos*
Hammond (1921)	None: slaughter-house sample	Sows	14.8	5	30–60	74	11.0
		Sows	16.7	4		71	11.9
		Sows	18.8	5		65	12.2
		Sows	22.6	5		54	12.2
Burger (1952)	None	Sows	13.3	12	26	74	9.8
		Sows	16.0	14		79	12.6
		Sows	19.6	14		66	13.0
Perry (1954)	None	Sows and gilts	12.7	26	Various	74	9.4
		Sows and gilts	16.2	24		71	11.5
		Sows and gilts	19.5	24		62	12.1
		Sows and gilts	25.0	25		58	14.4
Gibson *et al.* (1963)	Control	Gilts	10.7	13	25	82	8.8
	Super-ovulated	Gilts	23.8	5		50	11.8

Longnecker *et al.* (1965)	Controls	Sows	11.5	6	25	78	9.0
		Sows	23.9	9		66	15.8
Hunter (1966)	Super-ovulated	Gilts	35.4	8	25	48	17.0
Day and Longnecker (1968)	Controls	Gilts	12.2	28	26	76	9.3
	Super-ovulated	Gilts	23.4	42		57	13.3
Longnecker and Day (1968)	Controls	Sows	13.1	18	25	77	10.1
	Super-ovulated	Sows	25.1	19	40	61	15.3
Pope *et al.* (1968)	Controls	Gilts	12.8	2	30–40	82	10.5
	Super-ovulated	Gilts	14.5	3		83	12.0
	Super-ovulated	Gilts	19.7	5		71	14.0
Bazer *et al.* (1969)	Controls	Gilts	13.0	24	25	74	9.6
	Super-induction	Gilts	23.9	11		50	12.0

effect seems to be only applicable where ovulation rate is extreme. Some studies have in fact shown no relationship between the two (Perry, 1960; Pomeroy, 1960) unless the ovulation rate was exceptionally high. *Table 7.4* gives the results of a number of experiments relating ovulation rate to embryo survival and it can be seen that generally as ovulation rate increases, embryo survival decreases. Wrathall (1971) has concluded that embryo survival decreases by 1.24 per cent for every unit increase in ovulation rate.

Experiments using superovulation have in some cases significantly raised litter size at 25–40 days post coitum (*Table 7.4*) but there was also an elevated percentage loss compared to untreated controls (Hunter, 1966; Longnecker and Day, 1968; Pope, Vincent and Thrasher, 1968). Other workers have failed to observe any differences (Day *et al.*, 1967) suggesting that any eggs shed at ovulation above 'normal' limits are not represented later by viable embryos due to the limiting capacity determined by embryo survival or maternal limitation. This latter concept of a maternal ceiling proposed by Brooks (1970) is of particular relevance to the gilt which despite elevated ovulation rates due to exogenous applications will always have an upper limit to the number of eggs carried to 3 weeks post coitum and embryo mortality is elevated also to give a litter size determined by this maternal limitation.

INHERENT DEFECTS

A significant component of embryonic loss arises from genetic defects within the embryo itself and these lethal factors seem to be responsible for approximately 50 per cent of the losses incurred in the blastocyst stage (Wrathall, 1971). Harmful genes passed down from the parents make up a proportion of the defects as do genetic accidents such as mutations in the gametes both before and after release from the gonads (Bishop, 1964). Accidents at fertilization can also account for defects in the zygote and these include such phenomena as: point mutations, deletions, replications, invasions, translocations and polyploidy. These factors show a minimal expression in terms of losses in the early cleavage stages and the bulk of loss due to these lethal factors is probably from blastulation onwards and in particular at gastrulation (Briggs and King, 1961). McFeely (1967) has carried out chromosome analysis on day 10 pig blastocysts and has observed about 10 per cent of embryos at this stage showing cytogenetic abnormalities.

LOSSES RELATED TO MATERNAL EFFECTS

Another possibility for embryonic loss has been put forward by a research group working at the University of Florida. The basis for the hypothesis is that early embryos compete for some essential biochemical substance present in limited quantities in uterine secretions. Bazer *et al.* (1969) showed that by transferring embryos from one litter to another where the transferred embryos were younger by only a few hours than the indigenous embryos, the transfered embryos were unable to compete and showed poor survival. This is interpreted as meaning competition for some critical substance thereby providing an early natural selection system for vigorous embryos. Murray *et al.* (1971) subsequently demonstrated that the total protein of the uterine secretions begins to rise to a peak on day 15 of the oestrous cycle and decreases by day 17, or in other words, closely follows luteal activity. Stabenfeldt *et al.* (1969) have also reported that peak plasma progesterone concentration coincides with peak protein level in uterine secretions. Furthermore two specific uterine protein fractions have been identified which appear on days 9–16 and on days 12–16 of the oestrous cycle and it has been suggested (Knight, Bazer and Wallace, 1973a) that these are primarily progesterone induced. Knight, Bazer and Wallace (1973b) have also demonstrated the existence of a correlation between uterine protein and the number of corpora lutea. Despite these observations however, Knight, Bazer and Wallace (1973a, b) have failed to show a relationship between progesterone-induced uterine secretory activity and either embryo survival or number of live embryos despite the fact that a positive dose response relationship exists between progesterone and the protein fractions in the uterus. A relationship was observed, however, between progesterone-induced proteins and uterine weight, allantoic volume and placental development so further work may shed more light on the mode of action (if any) of uterine secretions on the survival of embryos.

THE EFFECTS OF STEROID HORMONE ADMINISTRATION

Ovarian hormone administration has been tried in a number of experiments as a means of augmenting embryo survival rates – *Table 7.5* gives the results from a number of such studies. Reddy, Mayer and Lasley (1958) observed a significant increase in the number of fetuses alive at day 55 post coitum when gilts were given small and balanced daily doses of progesterone plus oestrone by injection during early gestation. In contrast, other workers (Haines, Warnick and Wallace, 1958; Davis and Sorrensen, 1959; Spies *et al.*, 1959; Day, Romack and Lasley, 1963; Morrissette *et al.*, 1963) have failed to observe a consistent beneficial response as a result of the administration of ovarian hormones. Day, Romack and Lasley (1963) have suggested that hormone therapy may increase the survival rate only in those animals with an endogenous deficiency of steroid hormones and

Table 7.5. The effect of exogeneous progesterone and oestrogen on embryo survival

Author	*Treatment*	*Timing of dose* (days post coitum)	*Timing of slaughter* (days post coitum)	*Ovulation rate*	*Number of live embryos*	*Embryo survival* (%)
Reddy *et al.* (1958)	Progesterone, oestrogen 1000:1	4–10	55	–	0.9	78.3
	Progesterone, oestrogen 1000:1	14–24	55	–	8.7	81.3
	Progesterone, oestrogen 2000:1	4–10	55	–	9.0	81.9
	Progesterone, oestrogen 2000:1	14–24	55	–	10.7	86.5
	Control	–	55	–	7.9	76.7
Haines *et al.* (1958)	Progesterone injection	3–25	25	13.7	10.3	76.5
	Control	–	25	15.0	13.6	90.7
	Progesterone injection	3–25	40	13.0	9.0	68.6
	Control	–	40	13.1	11.3	86.0
Spies *et al.* (1959)	Progesterone	0–25	18	–	–	69
	Control	0–25	18	–	–	64
	Progesterone	0–25	18	–	–	81
	Control	0–25	18	–	–	86
Day *et al.* (1963)	Progesterone–oestrogen implant	11–25	25	15.9	11.6	73
	Control	11–25	25	16.4	11.0	67
	Progesterone–oestrogen implant	7–25	25	12.6	9.4	75
	Control		25	12.4	9.3	75
Morrissette *et al.* (1963)	Oral progesterone–oestrogen	0–25	25	18.3	16.3	89.1
	Control	–	25	20.7	12.8	62.1
	Oral progesterone–oestrogen	0–25	40	16.1	12.4	77.0
	Control	–	40	16.9	12.5	74.3

from their study they have reported a decrease in the proportion of animals showing very high embryo loss following hormone treatment.

UTERINE SPACE

At first sight a limitation imposed by available uterine space on the number of surviving embryos seems a plausable hypothesis. Perry (1954, 1959) has observed that the proportion of litters showing no losses whatsoever is greater than one might expect and the death of one embryo seems to contribute to the death of others close by. Aberrant embryo spacing might be related to this observation and where spacing is optimal, the influence of one embryo on a neighbour is minimized.

There is a great variation in uterine length and weight in early pregnancy (Perry and Rowlands, 1962; Dhindsa, Dziuk and Norton, 1967; Rigby, 1968; Varley and Cole, 1976) but no relationship has been established between uterine length, number of corpora lutea and number of surviving embryos (Varley, 1976). Available space in the uterus has been limited experimentally in a number of studies by either removal of whole horns or ligation of part of a uterine section. Dziuk (1968) and Fenton *et al.* (1970) have shown that the available uterine tissue or space does not control uterine capacity. The uterus is in fact capable of compensatory capacity under an unknown external stimulus. Dziuk (1968) has concluded from his studies that overcrowding of the uterus does not affect survival rates in early pregnancy but he has conceded that available space probably limits survival in the later stages of pregnancy after day 25 post coitum. Webel and Dziuk (1974) have substantiated this hypothesis and have observed that decreasing embryo space to half normal does not affect survival rates to day 30 but after this time fetal survival rates are significantly affected. A similar conclusion has been made by Pope *et al.* (1972) who have used an egg transfer technique to overcrowd the uterus artificially with developing embryos.

BACTERIAL INFECTION

The bacterial contamination of the uterus has been another area of investigation as a possible cause of some embryonic losses. Scofield (1969) found that approximately half observed sows and gilts had substantial uterine bacterial contamination, *Escherichia coli* and *Staphylococcus albus* being the commonest species found. Hajovsky and Gamcik (1966) however failed to identify a relationship between uterine infection and prenatal loss but other work shows, on the contrary, lower litter sizes and reduced conception rates resulting from bacterial infections (Evans, 1967). Scofield (1969) and Scofield,

Table 7.6. The effect of feeding antibiotics to female pigs on subsequent fertility

Author	*Treatment*	*Timing of dose*	*Time of estimation of fertility*	*Ovulation rate*	*Litter size*	*Embryo survival (%)*
Tribble *et al.* (1956)	40 mg/kg chlorotetracycline	Weaning to slaughter	25 days post coitum	10.9	9.8	90.0
	Control		25 days post coitum	11.7	9.3	80.8
Ruiz *et al.* (1966)	Chlorotetracycline + sulphamethazine penicillin	1 week before to 3 weeks post coitum	Full term	–	11.1	–
	Control		Full term	–	9.6	–
Ruiz *et al.* (1968)	Penicillin + chlorotetracycline	1 week before to 3 weeks post coitum	Full term	–	11.1	–
	Control		Full term	–	9.6	–
Sewell and Carmon (1958)	Chlorotetracycline	Weaning and throughout life	Full term	–	8.63	–
	Control		Full term	–	8.0	–
Fowler and Robertson (1954)	Chloromycetin	3 months of age to slaughter	25 days post coitum	10.9	8.6	81.0
	Control		25 days post coitum	11.6	8.0	68.5

Dean and Tribble (1962)	Chlorotetracycline	1 week before to 3 weeks post coitum	Full term	–	11.0	–
	Control		Full term	–	9.0	–
Davey *et al.* (1955)	Aureomycin 10 mg	Weaning to full term	Full term	–	10.0	–
	Aureomycin 50 mg	Weaning to full term	Full term	–	7.2	–
	Aureomycin 100 mg	Weaning to full term	Full term	–	6.7	–
Sosa *et al.* (1963)	Tylosin	Puberty to slaughter	26 days post coitum	15.4	12.9	83.9
	Control		26 days post coitum	14.7	13.2	89.6
	Tylosin	Puberty to slaughter	26 days post coitum	13.8	11.4	87.9
	Control		26 days post coitum	12.6	10.0	80.0
Mayrose *et al.* (1964)	Tylosin	1 week before to 3 weeks post coitum	Full term	–	12.7	–
	Control		Full term	–	11.9	–

Clegg and Lamming (1974) have looked at uterine infections in gilts at day 9 and day 13 post coitum. The presence of high level infection was still compatible with successful pregnancy but embryo survival was reduced significantly when infection was observed. Infection accounted for 40 per cent of ova lost between days 9 and 13 post coitum. Sources of infection arise mainly from the boars preputial fluid at mating or from semen itself both of which have a high bacterial content (Reed, 1969).

An attempt was made by Scofield (1972) to sterilize the vagina and uterus at mating by flushing out with antiseptic solution. This did not result in increased survival rates, however, as determined by litter size at full term.

The feeding of antibiotics to breeding females in the service period or prior to and at puberty has been attempted many times as a means of increasing fertility. Results are given in *Table 7.6* and, as can be seen, are very variable and no consistent response to antibiotic feeding has been observed.

In view of present limitations on the feeding of antibiotics to livestock in the UK and elsewhere, the technique if developed would probably be limited in application and indeed might represent a very real long term health hazard due to the possibility of bacterial immunity to antibiotics.

NUTRITION AND EMBRYO SURVIVAL

The relationships between nutrition and embryo survival have been well reviewed by a number of authors (Brooks, 1970; Scofield, 1969, 1972; Anderson and Melampy, 1972).

Specific dietary factors such as vitamins and minerals have the result of the loss of the whole litter when in a critically deficient state. On the other hand wide variations in the level and source of protein do not affect embryo mortality (Tassell, 1967).

Experiments on energy intake and its effect on embryo survival have been centred on energy intake either up to and just prior to mating or in the immediate post coitum period. Dutt and Chaney (1968) used three feed levels following mating which were 4.1, 2.4 or 1.2 kg feed/day (ME 51.2, 30.0, 15.0 MJ) and the results showed that as feed intake increased embryo mortality increased in gilts. This result was independent of pre-mating feed intake. Ray and McCarty (1965) found a similar effect in a more extreme situation when they fasted gilts for 24, 48 or 72 hours following mating. Fasting the gilts for the longer periods significantly improved embryo survival rates when compared with controls. In contrast to this finding, Schultz *et al.* (1965) fed levels of 1.8 or 2.7 kg feed/day following mating and found that the higher level significantly improved embryo survival. In

Table 7.7. Short term feeding influences on embryo survival

Author	*Feed level* (kg/day)	*Timing of treatment*	*Timing of slaughter* (days post coitum)	*Litter size*	*Embryo survival* (%)
Brooks (1970)	1.8	Puberty to slaughter	20	9.3	85.2
Experiment 1	3.6	Puberty to slaughter	20	10.0	84.4
	3.6	2 days post coitum	20	10.8	91.2
Brooks (1970)	1.8	Puberty to slaughter	20	10.7	94.2
Experiment 2	3.6	On day of mating	20	10.6	85.1
Brooks (1970)	1.8	During oestrus	25	13.7	85.3
Experiment 5	3.6	At oestrus	25	13.1	85:5

a later trial, however, the same authors found no significant differences in embryo survival for postmating feed levels of 1.8 or 3.6 kg/day.

It may be concluded, therefore, that while premating feed level needs to be high for a response in ovulation rate, postmating levels are probably best kept down to a minimum due to any possible detrimental effect of high energy intake on embryo survival.

The effects of plane of nutrition up to and during mating on embryo survival have been studied by Brooks (1970). A quantitative account is given in *Table 7.7* which shows the results of experiments where the effect of very short periods of high plane feeding at or around oestrus and ovulation has been evaluated as a means of increasing reproductive performance. Results are very inconsistent and *Table 7.7* shows that no real response in embryo survival is evident following this sort of treatment.

7.9 Fetal losses

Fetal losses from about day 30 post coitum onwards are smaller than embryo losses but are still significant and result from different mechanisms. Wrathall (1971) estimated this fetal loss to full term to be up to 10 per cent not including losses in the perinatal period.

Fetal losses are dependent on the number of surviving embryos entering the fetal period, and, as might be expected, the level of fetal mortality increases with the number entering this phase.

The biggest single influence on fetal loss seems to be available uterine space. Hammond (1921) recorded a trend towards a larger proportion of dead fetuses in the most crowded horns in the early fetal stages. Most workers have also found that a higher percentage of fetal death occurs when there are more than five fetuses per horn (Perry and Rowell, 1969). Perry and Rowell (1969) have also reported that when the number of fetuses per horn exceeds five, then those at the ends of the horn have an increasing body weight advantage over those in the middle. Fetuses at the ovarian end also tend to be heavier than those at the cervical end. No indication of a systematic location and mortality relationship existed, however, and deaths appeared to occur at random.

Overcrowding and competition for available endometrial surface are of considerable importance in fetal loss but when one death occurs, the dead fetus remains intact inside the uterus until parturition: how this affects the survivors is not known. Waldorf *et al.* (1957) have found no evidence to suggest that decomposition products from dead fetuses retard the growth and development of remaining viable fetuses but conversely there is no evidence either to suggest that growth of neighbouring fetuses is enhanced.

With respect to the effect of placental attachment area and fetal development, Rathnasbapathy *et al.* (1956) found a significant positive correlation between the body weight of fetuses and the uterine space occupied by 55 day old fetuses. The relationship was not a linear one but followed more closely a 'diminishing returns' law. Over and above 40 cm of uterine space per fetus additional space resulted in no further increase in fetal weight. This would imply that uterine length is a controlling factor in fetal survival.

Wrathall (1971) has pointed out that it is difficult to separate the effect of the fetuses on the uterus from the ability of the uterus to support more fetuses. However Pomeroy (1960) observed no increase in the length of the uterus beyond week 3 or 4 of pregnancy. It does seem highly likely therefore that the uterine length *per se* is the major factor affecting survival.

7.10 Conclusions

The period of gestation is probably the most important phase of a sow's reproductive life and in a high output situation she should be spending about two-thirds of her life pregnant.

The successful culmination of pregnancy is the delivery of a large number of viable piglets of satisfactory birthweight. There is a tremendous variability in sows (and gilts) in achieving this aim. A vast knowledge has already been gained of the mechanisms controlling the process from fertilization to parturition but quite clearly there is a lot more to be determined before we can influence and control the development of pregnancy. The objective in the future should be to identify those factors related to highly successful gestation and then to reproduce these factors in all sows and gilts to bring up the performance of the whole herd. Every pig farmer has records of the occasional sow which consistently produced 16 piglets at term and yet others struggle to produce 7s and 8s. Explanation for this almost certainly should transpire from a more precise understanding of embryonic and fetal survival mechanisms. Ovulation rate, although probably limiting reproductive output in gilts, does not seem to be a limiting factor to the sow where 15–20 eggs may be shed at one ovulation. The critical losses of embryos at blastulation and implantation and fetuses later on in pregnancy due to space limitations are the real and significant problems yet to be solved. Steroid balance and pathological status may be two of the most important areas worthy of further investigation and endocrine imbalances in relation to uterine secretory activity might eventually prove to be the key to the problem.

What is at least now known is a reasonable picture of optimal feed input in pregnancy to achieve maximum reproductive output and the work of the late Professor Frank Elsley and others in the late 1960s showed conclusively that high energy input in gestation is not necessary or even desirable to achieve high prolificacy. This topic is dealt with in depth in Chapter 8.

Another area which may be important to the successful gestation period is the environment of the sow and gilt socially and physically. Sow housing has radically changed its format in the last five years and where once group housing and semi-enclosed yards were the order of the day, a large proportion of sows are now housed individually in a stall or tether in a controlled environment house. There are indications that this latter system might be linked with such things as abortion, infertility and low fecundity but this has not been evaluated or confirmed. If there are systematic problems of these kinds then we ought to know more about how to keep them to a minimum or better to eradicate them completely. It is also possible that with the older housing systems the problems were still there but because of the difficulties in monitoring individual sows with such systems they simply went unnoticed.

7.11 References

ANDERSON, L.L. and MELAMPY, R.M. (1972). In *Pig Production* (Ed. by D.J.A. Cole), pp. 329–366. London, Butterworths

BAZER, F.W., ROBINSON, O.W., CLAWSON, A.J. and ULBERG, L.C. (1969). *J. Anim. Sci.* **29**, 30–31

BAZER, F.W., ROBINSON, O.W. and ULBERG, L.C. (1969). *J. Anim. Sci.* **28**, 145

BISHOP, M.W.H. (1964). *J. Reprod. Fert.* **7**, 383–396

BRAUDE, R., CLARKE, P.M. and MITCHELL, K.C. (1964). *J. agric. Sci., Camb.* **45**, 19–27

BRIGGS, R. and KING, T.J. (1961). In *The Cell : Biochemistry, Physiology and Morphology* (Ed. by J. Brackett and A.E. Mirsky), Volume 1, pp. 537–617. New York, Academic Press

BRINSTER, R.L. (1965a). *J. exp. Zool.* **158**, 59–67

BRINSTER, R.L. (1965b). In *Preimplantation Stages of Pregnancy* (Ed. by G.W. Wolstenholme and M. O'Connor), pp. 60–88. Churchill, London

BRINSTER, R.L. (1967). *Expl Cell Res.* **47**, 271–275

BRINSTER, R.L. (1968). *Expl Cell Res.* **51**, 330–335

BRINSTER, R.L. (1973). In *Handbook of Physiology, Section 7* (Ed. by S.R. Geiger). Washington, D.C., American Physiological Society

BRINSTER, R.L. (1974). *J. Anim. Sci.* **38**, 1003–1012

BROOKS, P.H. (1970). PhD Thesis. University of Nottingham

BURGER, J.F. (1952). *Onderstepoort. J. Vet. Res. Suppl.* **2**, 3–218

CORNER, G.W. (1921). *Contr. Embryol.* **13**, 119–146

CROMBIE, P.L. (1970). *J. Physiol., Lond.* **210**, 101–102

CROSS, M.A. and BRINSTER, R.L. (1969). *Expl Cell Res.* **58**, 125–129

DAVEY, R.J., GREEN, N.W. and STEVENSON, J.W. (1955). *J. Anim. Sci.* **14**, 507–512

DAVIS, W.F. and SORENSEN, A.M. (1959). *J. Anim. Sci.* **18**, 1549–1552

DAY, B.N., ANDERSON, L.L., HAZEL, L.N. and MELAMPY, R.M. (1959). *J. Anim. Sci.* **18**, 675–680

DAY, B.N. and LONGNECKER, D.E. (1968). *Proc. 6th int. Congr. Anim. Reprod., Paris,* p.267

DAY, B.N. and LONGNECKER, D.E., JAFFE, S.C., GIBSON, G.W. and LASLEY, J.F. (1967). *J. Anim. Sci.* **26**, 777

DAY, B.N., ROMACK, F.E. and LASLEY, J.F. (1963). *J. Anim. Sci.* **22**, 637–639

DEAN, B.T. and TRIBBLE, L.F. (1962). *J. Anim. Sci.* **21**, 207–209

DHINDSA, D.S., DZIUK, P.J. and NORTON, H.W. (1967). *Anat. Rec.* **159**, 325–330

DIEHL, J.R. (1974). *Diss. Abstr. Int. B.* **35(2)**, 611

DU MESNIL DU BUISSON, F. (1966). PhD Thesis. Paris

DU MESNIL DU BUISSON, F. and DAUZIER, L. (1957). *C.T. Séanc. Soc. Biol.* **151**, 311–313

DUTT, R.H. and CHANEY, C.H. (1968). *Prog. Rep. Ky Exp. Stn No. 176*

DZIUK, P.J. (1968). *J. Anim. Sci.* **27**, 673–676

DZIUK, P.J., POLGE, C. and ROWSON, L.E. (1964). *J. Anim. Sci.* **23**, 37–42

EDGERTON, L.A. and ERB, R.E. (1971). *J. Anim. Sci.* **32**, 515–524

ENGLISH, P. (1968). PhD Thesis. University of Aberdeen

EVANS, L.E. (1967). Cited in *J. Anim. Sci.* **27**, 1601

FENTON, F.R., BAZER, F.W., ROBISON, O.W. and ULBERG, L.C. (1970). *J. Anim. Sci.* **31**, 104–106

FOWLER, S.H. and ROBERTSON, G.L. (1954). *J. Anim. Sci.* **13**, 949–954

GIBSON, E.W., JAFFE, S.C., LASLEY, J.F. and DAY, B.N. (1963). *J. Anim. Sci.* **22**, 858

HAFEZ, E.S.E. (1976). *Reproduction in Farm Animals.* Philadelphia, Lea and Febiger

HAINES, C.E., WARNICK, A.C. and WALLACE, H.D. (1958). *J. Anim. Sci.* **17**, 879–885

HAJOVSKY, T. and GAMCIK, P. (1966). *Folia Vet.* **10**, 205–214

HAMMOND, J. (1921). *J. agric. Sci., Camb.* **11**, 337–366

HUNTER, R.H.F. (1966). *Anim. Prod.* **8**, 457–465

KING, J.W.B. and YOUNG, G.B. (1957). *J. agric. Sci., Camb.* **48**, 457–463

KNIGHT, J.W., BAZER, F.W. and WALLACE, H.D. (1973a). *J. Anim. Sci.* **36**, 61–65

KNIGHT, J.W., BAZER, F.W. and WALLACE, H.D. (1973b). *J. Anim. Sci.* **39**, 743–746

LIGGINS, G.C. (1972). *J. Reprod. Fert. Suppl.* **16**, 85

LONGNECKER, D.E. and DAY, B.N. (1968). *J. Anim. Sci.* **27**, 709–711

LONGNECKER, D.E., LASLEY, J.F. and DAY, B.N. (1965). *J. Anim. Sci.* **24**, 924

LUNAAS, T. (1962). *J. Reprod. Fert.* **4**, 13–20

LUTWAK-MANN, C. (1959). In *Implantation of Ova* (Ed. by P. Eckstein), pp. 35–46. Cambridge, Cambridge University Press

MAYROSE, V.B., SPEER, V.C., HAYS, V.W. and McCALL, J.T. (1964). *J. Anim. Sci.* **23**, 737–740

McFEELY, R.A. (1967). *J. Reprod. Fert.* **13**, 579–581

McLAREN, A. (1968). *Proc. R. Instn Gt Br.* **42**, 153–170

MELAMPY, R.M., HENDRICKS, D.M., ANDERSON, L.L., CHEN, C.L. and SCHULTZ, J.R. (1966). *Endocrinology* **78**, 801–804

METCALFE, J., BARTELS, H. and MOLL, W. (1967). *Physiol. Rev.* **47**, 782–838

MORRISSETTE, M.C., McDONALD, L.E., WHATLEY, J.A. and MORRISON, R.D. (1963). *Am. J. Vet. Res.* **24**, 317–323

MURRAY, F.A.Jr., BAZER, F.W., RUNDELL, J.W., VINCENT, C.K., WALLACE, H.D. and WARNICK, A.C. (1971). *J. Reprod. Fert.* **24**, 445–448

MURRAY, F.A.Jr., BAZER, F.W., WARNICK, A.C. and WALLACE, H.D. (1971). *J. Anim. Sci.* **32**, 388

PARLOW, A.F., ANDERSON, L.L. and MELAMPY, R.M. (1964). *Endocrinology* **75**, 365–376

PATTEN, B.M. (1953). In *Embryology of the Pig* (Ed. by B.M. Patten). London, McGraw

PERRY, J.S. (1954). *J. Embryol. exp. Morph.* **2**, 308–322

PERRY, J.S. (1959). *Int. J. Fert.* **4**, 142–150

PERRY, J.S. (1960). *J. Reprod. Fert.* **1**, 71–83

PERRY, J.S. and ROWELL, J.C. (1969). *J. Reprod. Fert.* **19**, 527–534

PERRY, J.S. and ROWLANDS, I.W. (1962). *J. Reprod. Fert.* **4**, 175–188

PHILLIPPO, M. (1967). In *Proceedings of a Symposium on Nutrition of Sows, Nottingham*, pp. 6–10. London, Pig Industry Development Authority

POLGE, C. and DZIUK, P.J. (1970). *J. Anim. Sci.* **31**, 565–566

POMEROY, R.W. (1960). *J. agric. Sci., Camb.* **54**, 31–56

POPE, C.E., CHRISTENSON, R.K., ZIMMERMAN, D.R. and DAY, B.W. (1972). *J. Anim. Sci.* **35**, 805–810

POPE, C.E., VINCENT, C.K. and THRASHER, D.M. (1968). *J. Anim. Sci.* **27**, 303

RAESIDE, J.I. (1963a). *J. Reprod. Fert.* **6**, 421–426

RAESIDE, J.I. (1963b). *J. Reprod. Fert.* **6**, 427–431

RATHNASBAPATHY, V., LASLEY, J.F. and MAYER, D.T. (1956). *Res. Bull. Mo. agric. Exp. Stn* **615**

RAY, D.E. and McCARTY, J.W. (1965). *J. Anim. Sci.* **24**, 660–663

REDDY, V.B., MAYER, D.T. and LASLEY, J.F. (1958). *Res. Bull. Mo. agric. Exp. Stn* **667**

REED, H.C.B. (1969). *Br. vet. J.* **125**, 272–280

RIGBY, J.P. (1968). *Res. vet. Sci.* **9**, 551–556

ROMBAUTS, P., FEVRE, J. and TARQUI, M. (1971). *Journées de la recherche porcine en France*, pp. 173–178. Paris, L'Institut Technique du Porc

RUIZ, M.E., SPEER, V.C. and HAYS, V.W. (1966). *J. Anim. Sci.* **25**, 1250

RUIZ, M.E., SPEER, V.C., HAYS, V.W. and SWITZER, W.P. (1968). *J. Anim. Sci.* **27**, 1602–1607

SCHULTZ, J.R., SPEER, V.C., HAYS, V.W. and MELAMPY, R.M. (1965). *J. Anim. Sci.* **24**, 929

SCOFIELD, A.M. (1969). PhD Thesis. University of Nottingham

SCOFIELD, A.M. (1972). In *Pig Production* (Ed. by D.J.A. Cole), pp. 367–384. London, Butterworths

SCOFIELD, A.M., CLEGG, F.G. and LAMMING, G.E. (1974). *J. Reprod. Fert.* **36**, 353–361

SEWELL, R.F. and CARMON, J.L. (1958). *J. Anim. Sci.* **17**, 752–757

SHEARER, I.J., PURVIS, K., JENKIN, G. and HAYNES, N.B. (1972). *J. Reprod. Fert.* **30**, 347–360

SHORT, R.E. (1960). *J. Reprod. Fert.* **1**, 61–70

SOSA, E., WALLACE, H.D., WARNICK, A.C. and COMBS, G.E. (1963). *J. Anim. Sci.* **22**, 745–751

SPIES, H.G., ZIMMERMANN, D.R., SELF, H.L. and CASIDA, L.E. (1959). *J. Anim. Sci.* **18**, 163–172

STABENFELDT, G.H., AKINS, E.L., EWING, L.L. and MORRISSETTE, M.G. (1969). *J. Reprod. Fert* **20**, 443–447

STERN, S. and BIGGERS, J.D. (1968). *J. exp. Zool.* **168**, 61–66

TASSELL, R. (1967). *Br. vet. J.* **123**, 76–83

THOMPSON, J.L. and BRINSTER, R.L. (1966). *Anat. Rec.* **155**, 97–103

TRIBBLE, L.G., PEANDER, W.H., LASLEY, J.F., ZOBRISKY, S.E. and BRADY, A.J. (1956). *Res. Bull. Mo. agric. Exp. Stn* **609**

VARLEY, M.A. (1976). PhD Thesis. University of Nottingham

VARLEY, M.A. and COLE, D.J.A. (1976). *Anim. Prod.* **22**, 79–85

WALDORF, D.P., FOOTE, W.C., SELF, H.L., CHAPMAN, A.B. and CASIDA, L.E. (1957). *J. Anim. Sci.* **16**, 976–985

WARNICK, A.C., GRUMMER, R.H. and CASIDA, L.E. (1949). *J. Anim. Sci.* **8**, 569–577

WARWICK, B.L. (1926). *Anat. Rec.* **39**, 29–33

WEBEL, S.K. and DZIUK, P.J. (1974). *J. Anim. Sci.* **38**, 960–963

WRATHALL, A.E. (1971). *Prenatal Survival in Pigs.* Slough, England, Commonwealth Agriculture Bureau

Chapter 8

Litter size

One major purpose of a pig breeding enterprise is to produce offspring which are subsequently grown and sold. The output from a herd is measured by the number of pigs sold/sow/year, calculated at weaning for breeding herds only and at slaughter for combined breeding/fattening units. Since litter size is the end-product of all the effort and manipulation that goes into a breeding herd, it is important to consider how it may be maximized.

First, the present situation in respect to the commercial output of weaner pigs will be considered. *Table 8.1* provides a summary of the more recent information relating to litter size at both birth and weaning for herds in the UK. This data demonstrates that, although litter size is reasonably high at birth, the actual number of piglets in each litter that are reared to weaning age and beyond is low. It should also be noted that the mortality figure of 17.3 per cent presented in *Table 8.1* is rather low, compared to the observations made by Phillips and Zeller (1941), working in the United States, who have reported that piglet mortality from birth to weaning averaged 22.9 per cent.

Table 8.1. Some estimates of litter size in UK herds

	Average
Litter size at birth – alive	10.4
dead	0.6
total	11.0
Litter size at weaning	8.6
Piglet mortality (%) from birth to weaning	17.3

Data compiled from Thomas and Burnside (1973), Ridgeon (1973, 1974), M.L.C. (1976).

What, then, may be done to improve litter size at weaning and hence overall sow productivity? In order to answer this question it is necessary to consider the stages at which the losses of potential weaner pigs occur. Furthermore, the relative contribution of each of these stages to the overall losses must be assessed.

8.1 Components of litter size

The production of a litter of pigs obviously begins with the release of eggs from the ovaries, continues through fertilization and conception, and terminates at the end of a pregnancy lasting approximately 115 days. Subsequently, the piglets have to survive the birth process and remain viable during the difficult period up to weaning. Within this framework there are many stages at which losses can occur. These are summarized below:

(1) Litter size potential is determined by ovulation rate.
(2) Losses occur due to failure of fertilization.
(3) Loss of embryos and fetuses during pregnancy.
(4) Losses at parturition.
(5) Losses from birth to weaning.

Each of these stages will now be considered in turn in order to determine their relative influence on the reproductive output of the breeding herd (measured by the number of weaner pigs produced).

OVULATION RATE

Since the number of ova shed provides the upper limit of litter size, it might be expected that ovulation rate would be the primary factor limiting the size of litter produced. However, this is not the case. In fact, with the possible exception of the gilt, it seems likely that ovulation rate will be in excess of requirements (*see* Chapter 5). Thus, the sow sheds more ova than she is capable of maintaining as viable embryos through a pregnancy. In addition, the adoption of simple management procedures (outlined in Chapter 5) can considerably reduce any limitation to litter size occasioned by low ovulation rates. Hence there is no reason why ovulation rate should limit the size of litter produced by either the gilt or the sow.

FERTILIZATION

The percentage of ova which are fertilized will have an effect on potential litter size of the animal. In practice fertilization rates are high (90–100 per cent) and show relatively little variation (*see* Chapter 6). Thus, although potential litter size is likely to be slightly reduced by losses at the fertilization stage, it seems unlikely that fertilization failure (except where this is total) will have a major effect on the size of litter that will ultimately be produced.

GESTATION

It is clear that the limitations to litter size provided by ovulation and fertilization rates are unlikely to be severe. In contrast, losses during pregnancy will drastically reduce the size of litter which will eventually be produced. Indeed, under normal conditions 20–40 per cent of all fertilized ova will be lost during the course of a pregnancy (*see* Chapter 7). Furthermore, as ovulation rate increases so, too, does the rate of embryo loss (*Table 8.2*). Thus, assuming ovulation rate is not

Table 8.2. The relationship between ovulation rate and embryo survival in the pig

Average ovulation rate	*Percentage of ova which became*			*Number of normal embryos*
	Normal fetuses	*Atrophic fetuses*	*Missing ova*	
22.6	54.0	14.2	31.9	12.20
18.8	64.9	14.9	20.2	12.20
16.7	71.6	16.4	11.9	11.96
14.8	74.3	10.8	14.9	10.99

From Hammond (1921).

limiting litter size, the employment of methods of increasing ovulation rate is unlikely to result in the birth of larger litters since a concomitant rise in embryo mortality will occur. This is clearly demonstrated in *Table 8.2*.

PARTURITION

Losses of piglets during the birth process are normally low, although they may be significantly increased in cases where piglet birth weights are in excess of 1.5 kg (Fahmy *et al.*, 1978). In general, losses at parturition are reported to account for approximately 3–5 per cent of all piglets born (Phillips and Zeller, 1941; Fahmy *et al.*, 1978). This figure for losses at parturition will include not only those piglets that are lost as a result of farrowing difficulties but also piglets which died during the final stages of pregnancy and were therefore born dead.

BIRTH TO WEANING

In *Table 8.1* it is stated that, on average, 17.3 per cent of all piglets born alive are lost during the period to weaning. This figure is, of course, subject to considerable variation (*see Figure 8.1*), although it is clear that an average figure of 15–20 per cent is common. The pattern of piglet loss during this period is shown in *Table 8.3*. This

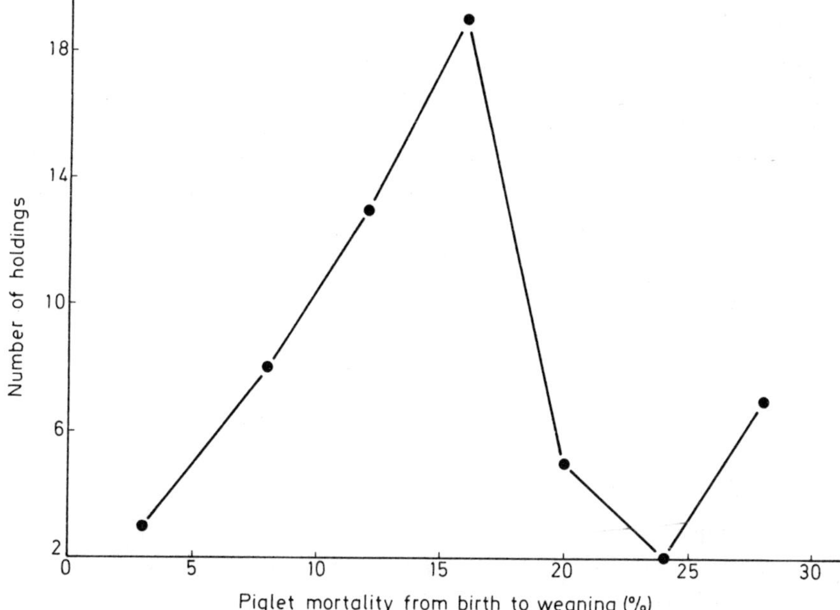

Figure 8.1 A typical range of pre-weaning piglet mortality rates (from Thomas and Burnside, 1973)

demonstrates that most piglet deaths occur during the first week after birth (68 per cent). Furthermore, the chances of piglets dying appear to diminish as they grow older, and they may be considered minimal post weaning (usually in the region of 1–2 per cent of weaned pigs are lost prior to attaining slaughter weight).

Table 8.3. The pattern of pre-weaning piglet deaths

Days after birth	*Percentage of total pre-weaning deaths*
1–2	50
3–7	18
8–21	17
22–56	15

Adapted from Veterinary Investigation Service (1960).

To summarize, there are two stages at which the major losses of potential weaner pigs occur. These are during pregnancy and in the period from birth to weaning. Under normal circumstances the effects of ovulation and fertilization rates are of a secondary nature, and are not considered to be major influences on litter size. Finally, parturition losses are generally low although as will be seen below, these may well be further reduced.

Some estimates of the overall losses occurring between ovulation and weaning are given in *Table 8.4*.

Table 8.4. Estimates of the fate of ova shed by sows during heats at which they were mated

Fate of ova	*Authors*	
	Phillips and Zeller (1941)	*Various**
Loss during pregnancy	31.3%	30%
Loss at parturition	3.2%	4%
Loss from birth to weaning	22.9%	17.3%

* This data is obtained from Wrathall (1971), Thomas and Burnside (1973), Ridgeon (1973, 1974), M.L.C. (1976), Fahmy *et al.* (1978).

8.2 Optimization of litter size

It has previously been quoted that the average number of live pigs/litter born in UK breeding herds is approximately 10.4, and this is reduced to 8.6 pigs/litter by the time of weaning. This level of output is not particularly high when one considers that the potential litter size of the sow at the time of ovulation is 15–25. The sow undoubtedly is potentially capable of rearing litters of 12–14 piglets.

What is the optimum litter size and how may this be achieved in practice? Ridgeon (1974) suggested a series of standards for litter size at birth and weaning. These suggestions (*Table 8.5*) indicate that a target litter size at birth should be 11–12 live piglets, with subsequent losses to weaning of no more than 12–13 per cent (i.e. 9.6–10.5 piglets weaned/litter). These targets are undoubtedly attainable and, if reached, could increase the annual productivity of a sow by at least three weaners compared to the national average.

Table 8.5. Suggested standards for litter size at birth and weaning

	Poor	*Average*	*Good*	*Best*
Litter size at birth (alive)	9.5	10.2	11.0	12.0
Litter size at weaning	7.8	8.7	9.6	10.5

From Ridgeon (1974).

To achieve these increases in output it is necessary to reduce both pre- and postnatal piglet losses. It has already been emphasized that these losses are primarily due to embryonic mortality and piglet deaths within the first week following birth. However, before discussing these two stages in detail it is worth reiterating that litter size may also be reduced, although usually to a lesser extent, by low ovulation and conception rates. To ensure that this does not occur it is recommended that the management procedures outlined in Chapters 5 and 6 are followed.

REDUCING LOSSES DURING PREGNANCY

High rates of embryo loss during pregnancy are quite common, although it is often difficult to ascertain the reasons for their occurrence. The information presented in Chapter 7 demonstrates that our present knowledge of the control of events during pregnancy is severely lacking. Hence, it is difficult to identify causes of embryo loss or, alternatively, to prescribe methods whereby embryo loss may be minimized. However, two areas of management have been identified as influences on embryo survival, these being the nutrition of the sow and the environment in which she is kept.

Generally, high levels of food or energy intake in early pregnancy have been linked with high rates of embryonic mortality (*see* Chapter 7), although it should be added that results on this relationship are not always consistent. It seems likely, therefore, that the feeding of low levels of energy during the first few weeks of pregnancy may minimize losses over this period. There appears to be little or no effect of food or energy intake on embryo survival during the remainder of gestation (higher energy intakes in late gestation will, however, increase piglet birth weights).

Increases in embryo mortality have also been reported to occur in response to adverse environmental conditions. In Chapter 7 it was suggested that housing sows individually, in either stalls or tethers, may result in adverse effects on the pregnant female. This is a general observation which, although possibly true, must be treated cautiously until the data are available either to substantiate or refute it. However, it is known that extremes of temperature and severe stress may influence embryo survival (Omtvedt *et al.*, 1971) and it is possible that these effects may be manifested to different extents in group- and individually housed sows. It is clear that individually housed sows will be less able to combat variations in temperature than will their group-housed counterparts. Furthermore, most sows appear to undergo a period of agitation and stress when first put into stalls or tethers. Such stress is likely to have an adverse effect on embryo survival, although it is questionable whether or not the stress will be any greater than that imposed when mixing sows for group housing.

REDUCING LOSSES FROM BIRTH TO WEANING

Table 8.4 shows that piglet losses at birth were only 3–4 per cent, but that subsequent losses to weaning were in the region of 15–25 per cent. Many producers might dispute the first of these figures (i.e. losses at birth) since reference to their litter records would suggest a much higher figure for the percentage of piglets born dead. However, here the distinction must be made between those piglets which are truly born dead and those which die during the first 24 hours. Piglets which are born dead are those which have either died during the final stages of pregnancy or have failed to survive the birth process.

In order to reduce fetal losses during late pregnancy it is again necessary to adhere to the suggestions made in the previous section regarding sow management during gestation. This may also be true for losses at birth, since these are mainly due to farrowing difficulties associated with heavy piglets. Thus, overfeeding the sow in late pregnancy results in heavy piglets at birth and increases the probability of losses during the birth process. However, as will be seen later, underfeeding pregnant sows may result in very low piglet birth weights which, in turn, reduces piglet viability during the first few days of life. Therefore, a compromise must be made when feeding the pregnant sow, such that she receives sufficient food to ensure reasonable piglet birth weights (1.3–1.5 kg) but does not produce overweight piglets (1.6 kg +) with their attendant farrowing problems.

Obviously piglet losses at birth may also be reduced if the stockman is present when the sow is actually farrowing since he will be able to give assistance with the births if needed. The use of prostaglandins to induce farrowing (*see* Chapter 7) may therefore have a role in this context. Indeed, inducing parturition at specific times allows for closer supervision of farrowings and should result in a reduction in piglet losses during the birth process.

Subsequent to birth piglet losses may occur as a result of many endogenous and exogenous factors. A typical pattern of these losses is shown in *Table 8.6.* Within the categories for piglet loss shown in this table it is clear that further generalizations may be made. Major sources of piglet deaths are:

(1) Inherent defects.
(2) Bacterial infections.
(3) Nutritional problems.
(4) Overlying and starvation.

Piglets which have inherent defects such as atresia ani or hydrocephalus will die soon after birth. Obviously there is little that can be done to prevent such losses apart from rigorous selection against them. In addition it seems likely that their occurrence will be more frequent with inbreeding. However, assuming that inbreeding has been avoided, it must be said that these losses have to be accepted.

Table 8.6. The causes and timing of pre-weaning mortality of piglets

Cause of death (%)	*Time post partum* (days)				
	<3	3–7	8–21	22–56	*TOTAL*
Overlying and starvation	33.8	8.5	6.0	1.9	50.2
Genetic defects	2.4	0.8	0.7	0.6	4.5
Respiratory problems	0.2	0.4	0.6	2.2	3.4
Alimentary problems	2.1	1.8	2.3	1.8	8.0
Deficiencies	0.3	0.4	1.5	1.5	3.7
Bacterial infections	2.6	2.2	2.5	3.8	11.1
Miscellaneous	0.8	0.9	1.2	1.7	4.6
No cause found	7.8	3.0	2.2	1.5	14.5
TOTAL	50.0	18.0	17.0	15.0	100.0

Adapted from Veterinary Investigation Service (1960).

Loss of piglets as a result of bacterial infection (most commonly *E.coli*) is rather different. Where such losses are apparent it is clear that a breakdown in hygiene has occurred, or that an inadequacy of either nutrition or environment has predisposed the piglets to infection. *Table 8.6* demonstrates that piglets are equally susceptible to such infection throughout the first 8 weeks of life. Hence, a routine of cleaning, disinfecting and resting of pens in which the young piglets are to be housed is of vital importance if losses due to infection are to be minimized. Regular inspection of litters and the early treatment of infected animals should reduce piglet losses still further. In cases where piglet viability has already been reduced by poor nutrition or a cold and/or wet environment, susceptibility to bacterial infection will clearly increase. Such a situation may be avoided by ensuring that the environment is optimum for the young piglets, that the sow is producing adequate milk, and that creep feed is introduced to the litter at an early age (usually at 2 weeks or before).

Piglet deaths due to nutritional inadequacies are common although, since weakened piglets often succumb to bacterial infection (*see above*), they are often recorded as deaths due to infection. At their simplest, nutritional problems are threefold; overfeeding, underfeeding and deficiencies of specific nutrients. The usual outcome of overfeeding is that piglets begin to scour, and this will increase the susceptibility of the young animal to bacterial infection. Underfeeding, due to either an inadequate supply of milk from the sow or the failure to provide sufficient creep feed, will also frequently result in bacterial infection, although in this case it is due to the low nutrient intake reducing piglet viability and, therefore, disease resistance. Deficiencies of specific nutrients are rarer, the most common deficient nutrient being iron. However, since it is now standard practice to provide this mineral in an exogenous form to the piglet soon after birth the losses due to such deficiencies should be minimal.

Inherent genetic defects, together with nutritional inadequacies and bacterial infection, clearly make a significant contribution to the rate of piglet loss between birth and weaning. However, by far the largest category of piglet loss during this period is that of overlying and starvation. Over 50 per cent of all losses between birth and weaning may be ascribed to this category, two-thirds of these deaths occurring in the first 2 days post partum. Losses due to starvation (i.e. failure to provide a warm, dry environment) may readily be avoided by the provision of an easily accessible creep area for the piglets. Such an area should be equipped with a heat lamp and bedding material (e.g. straw or wood shavings). Unfortunately, piglet deaths caused by overlying by the sow are not so readily avoided. They may, to a certain extent, be reduced by using a well designed farrowing crate and providing a creep area which will allow the litter to lie away from the sow. Also, since heavy sows tend to be the clumsiest, it would seem wise to prevent sows from becoming overfat. Finally, the use of breeds which are recognized to have good mothering ability may also help to reduce losses caused by overlying. However, it has to be accepted that some losses will occur as a result of overlying by the sow, although adherence to the points given above should considerably reduce such losses.

8.3 General considerations

At this point it is necessary to note several additional factors which may influence either the size of litter produced or the viability of the piglets in the post-partum period. The effects of early weaning on subsequent litter size are discussed in Chapter 9. However, it should be noted here that weaning sows earlier than 3 weeks does tend to result in a reduction in the size of litter subsequently produced (*see Table 9.6, Figure 9.9*). Indeed, until such time as the problem of increased embryonic mortality occasioned by early weaning is understood, it is clear that lower litter sizes must be accepted as an integral part of very early weaning systems. However, of equal importance are the possible influences of early weaning on the viability of the litter being weaned. Recent advances in the design of houses for early weaned piglets have undoubtedly resulted in considerable reductions in losses subsequent to weaning. The nutritional and environmental requirements of early weaned piglets are now well understood, as is the need for strict hygiene. Indeed, since all aspects of the young piglet's life can be more easily controlled once it has been removed from its dam, it might be expected that early weaning would result in a reduced rate of piglet loss during the post-partum period. However, *Table 8.6* shows that most piglet losses (approximately 68 per cent) occur in the first week of life, this normally being a time at which even litters that are to be early weaned are still with the sow.

Four other factors which may influence the rate of piglet mortality from birth to weaning are the number of piglets in the litter, variation of piglet size within the litter, piglet birth weight and the birth order of the piglets. Generally speaking, the larger the litter the greater is the expected mortality to weaning. This is certainly true for large litters (12+), although it is unlikely to be apparent in smaller litters. The reasons for increased losses in larger litters are that both the individual birth weights and the milk supply/piglet tend to be low, resulting in decreased piglet viability. However, a factor which will probably exert a more significant influence on piglet mortality is the variability of birth weights within a litter. Indeed, in litters where there are both large and small piglets the competition for food frequently results in the smaller piglets receiving the least food, and eventually dying. It should be emphasized that such size variations do tend to occur in every litter as a result of uterine conditions (a situation which becomes more acute as the parity of the sow increases). The smallest piglets come from the middle section of the uterine horns, with larger piglets being produced at the ovarian and cervical ends. These differences are accentuated following birth since those piglets which are born first (i.e. the larger piglets, from the cervical end of the uterine horns) are usually at the top of the teat order. This means that these piglets get the anterior teats which produce most milk. Therefore, the size differentials within the litter which are present at the time of birth tend to be continued, and even increased, during the lactation. This may explain why large size differences occur in piglets within large litters by the time of weaning. In contrast, the situation would not arise in smaller litters since there would always be surplus teats from which the smaller members of a litter could obtain their nutrient supply.

8.4 Conclusions

It has previously been stated that, under normal systems of weaning (i.e. 3 weeks and above), a target of 11–12 piglets born alive per litter should be aimed for. In the case of the gilt this target should be reduced to about 9–10 piglets/litter at birth. The achievement of these targets is dependent on initially attaining a high ovulation rate, and subsequently minimizing embryonic/fetal mortality during pregnancy.

To ensure that ovulation rate does not limit litter size in the gilt it is necessary to either delay mating until third oestrus or later, or to 'flush' the gilt for one oestrous cycle prior to mating at second heat (*see* Chapter 5). In the sow, ovulation rate will normally be in excess of requirements and hence is unlikely to limit litter size. However, it is recommended that high level feeding (3–4 kg/day) is employed for the short period from weaning to remating in order to ensure that no limitation occurs.

It is more difficult to prescribe methods whereby embryonic/fetal losses during pregnancy may be prevented or minimized since our knowledge of this period is severely limited. However, it is clear that pregnancy losses may be reduced by providing the pregnant sow with an optimum climatic, social and nutritional environment. The housing conditions of the sow should be such that she is not exposed to wide variations in ambient temperature during early or late pregnancy, and stress is kept to a minimum. Nutritionally, the evidence indicates that low level feeding (1.75–2.50 kg/day depending on sow size and condition) during the initial 3–4 weeks of pregnancy should reduce embryo mortality over this period. Subsequently, feed level during the remainder of pregnancy does not appear to exert any influence on the rate of survival of the embryos/fetuses. However, it should be noted that high level feeding over this period will result in heavy piglets which may cause problems, and possibly piglet losses, at the time of farrowing. In addition, sows which become overfat during this period will tend to be clumsy post partum, and therefore the probability of piglet deaths due to overlying is increased.

Losses at parturition are mainly due to farrowing difficulties associated with heavy piglets. However, most recorded losses at farrowing will also include piglet deaths within the first 12–24 hours post partum. In this category losses due to overlying by the sow and piglet starvation are the most common. To minimize such losses two factors must be born in mind; first, the design of the farrowing crate and creep area should be such as to provide maximum protection for the piglets, and secondly the supervision, or frequent inspection, of sows at, and immediately following, parturition should further reduce piglet deaths associated with overlying and starvation.

Once the live litter has been produced the aim should be to maintain as many viable piglets as possible through to weaning. For conventional weaning systems a target of no more than 12–13 per cent piglet mortality during lactation is suggested, this providing a litter size at weaning of 9.6–10.5 piglets. To achieve this target, piglet deaths due to overlying by the sow, starvation, nutritional inadequacies and bacterial infection must be minimized. In order to reduce losses due to overlying and starvation the two points given above relating to farrowing house design and inspection of sows and litters are particularly important. Losses due to malnutrition or undernutrition may be reduced by following the points outlined below:

(1) Ensure that the sow's milk supply is adequate for her litter.
(2) Provide all piglets with exogenous iron within the first few days following birth.

(3) Introduce small amounts of fresh creep feed daily, beginning when the litter is 10–14 days old.
(4) Do not overfeed the young piglets.

Finally, piglet deaths as a result of bacterial infection (particularly *E.coli*) may be minimized by strict hygiene. This involves the cleaning, disinfecting and resting of all pens prior to occupation by the young piglets. In addition, the prompt treatment of infected litters is necessary to reduce the mortality rate, although growth performance will obviously be retarded.

8.5 References

FAHMY, M.H., HOLTZMANN, W.B., MACINTYRE, T.M. and MOXLEY, J.E. (1978). *Anim. Prod.* **26**, 277
HAMMOND, J. (1921). *J. agric. Sci., Camb.* **11**, 337
MEAT AND LIVESTOCK COMMISSION (M.L.C.). (1976). *Feed Recording Results,* 1976
OMTVEDT, I.T., NELSON, R.E., EDWARDS, R.L., STEPHENS, D.F. and TURMAN, E.J. (1971). *J. Anim. Sci.* **32**, 312
PHILLIPS, R.W. and ZELLER, J.H. (1941). *Am. J. Vet. Res.* **5**, 439
RIDGEON, R.F. (1973). *Pig Management Scheme Results for 1973,* University of Cambridge
RIDGEON, R.F. (1974). *Pig Management Scheme Results for 1974,* University of Cambridge
THOMAS, W.J.K. and BURNSIDE, E. (1973). *Pig Production: Results of a Study in South West England in 1972–73.* University of Exeter
VETERINARY INVESTIGATION SERVICE (1960). *Vet. Rec.* **72**, 1240
WRATHALL, A.E. (1971). *Commonw. Agric. Bur., Commonw. Bur. Anim. Hlth Rev. Ser.* **9**

Chapter 9

Lactation

After parturition the sow or gilt enters the lactational phase. During this period all the metabolic and physiological systems are operating for the production of sufficient milk to rear the litter of piglets. Hence feed requirements in lactation are very much different compared to the sow in pregnancy. The objective must be the maintenance of a high growth rate from as many piglets as possible eventually leading to maximum weaning weight. The aim also is for the sow or gilt to rear as many of the piglets born alive to weaning or in other words to minimize piglet mortality. At the present time average losses in the nursing period may be around 12–15 per cent of all piglets born alive. Even on the best pig farms the figures occasionally rise much higher than this for a variety of reasons.

The traditional length of the lactation period has been 56 days. Weaning piglets after this time period means few problems with the subsequent nutrition of the piglets and equally few problems with the subsequent reproductive performance of the sow. Economic pressures have been responsible for the producers steadily reducing their lactation length in search of a higher reproductive output. Although the 8 week weaning system is good for both piglet and dam, this limits the farrowing index to 1.8–2.0 litters/sow/year. With a 3 week weaning system, 2.5 litters/sow/year can be produced at very little extra cost. In fact there is now a wide variety of different weaning systems being advocated but not all of them are satisfactory in terms of realized reproduction output. We shall look into this area in detail in the present chapter.

9.1 Physiology of lactation

The udder of the sow extends over the whole abdominal wall and the number of glands and teats can range from four to nine pairs. The modal number of pairs is five or six. Obviously a sow or gilt with only four pairs of teats has not the capacity to rear large litters of piglets to weaning even if a genetic ability to produce large litters at birth exists. Producers and breeders tend to apply considerable selection pressure to this characteristic and any potential breeding female with less than six pairs of anatomically sound teats is usually culled from groups of breeding gilts before any other trait is considered.

Each teat is traversed by two canals and each of these canals leads to a small dilation of the sinus and eventually ramifies into its own section of alveolar–lobular tissue. In other words there are two ducts opening into each teat and each of these individual ducts has its own self-contained duct and glandular system.

THE DEVELOPMENT OF THE GLAND SYSTEM

An illustration of mammary development for a 30 day old pig embryo is given in *Figure 9.1*. Mammary tissue is derived from the ectoderm in the embryo and differentiation of the eventual udder

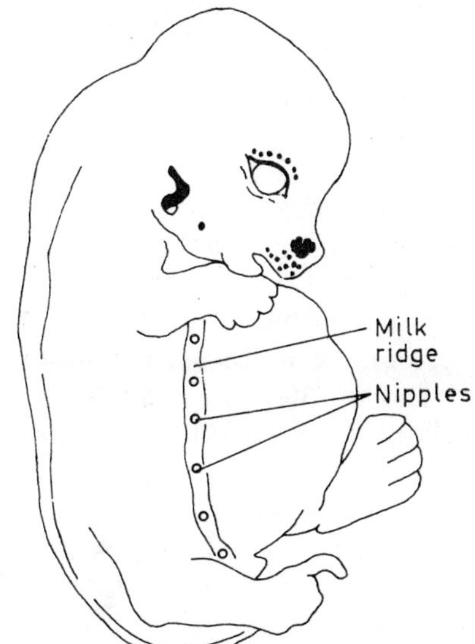

Figure 9.1 Drawing (X515) of 20 mm pig embryo showing milk ridge (from Frandson, 1974)

first becomes apparent in the very early embryonic stage when two parallel ridges of ectoderm appear which are known as 'milk lines'. Nodules along these milk lines form themselves into so-called mammary buds each of which is the progenitor of a teat and gland system.

At birth there is still relatively little development of the duct system but there is considerable differentiation of tissue into the stromal tissue of the udder.

By the time the female pig is 5 months old (i.e. just prior to or at puberty) the duct system is well developed, particularly in the regions proximal to the gland cisterns. Lobulo–alveolar development is subsequently present by day 45 post coitum and by day 60 of pregnancy the alveoli are at an advanced state of development. By this stage a very small quantity of secretory activity may be observed within the alveoli but they do not become distended as a result of this. Even 9 days prior to parturition the alveoli are still quite small.

Four days before parturition the alveoli begin to expand and fat globules may be seen in the lumen of the alveoli at about 2 days prior to parturition (Cross, Goodwin and Silver, 1958). Sows let their milk down some 12–24 hours prior to parturition and in fact this sign is often used as a sure indication of imminent labour.

HORMONAL CONTROL OF MAMMARY DEVELOPMENT

It would appear that a complex of hormones is involved in the control of mammary development. Oestrogens, progesterone, growth hormone, prolactin and adrenocortical hormones are all certainly implicated in this but the precise role of each has not yet been evaluated. In addition most of this work has been carried out in laboratory rodents and how far it can be extrapolated to the pig is not known.

LACTATION

Milk production is under the control of a number of hormones of which prolactin, ACTH and growth hormone are probably the most important (Kuhn, 1970). A peak of milk secretion and production occurs at or around week 3 of lactation after which milk synthesis and supply gradually fall off as the alveoli begin to regress in an orderly fashion. The pattern of milk yield in the sow is shown in *Figure 9.2* with peak yield reaching a daily output of just over 7 kg.

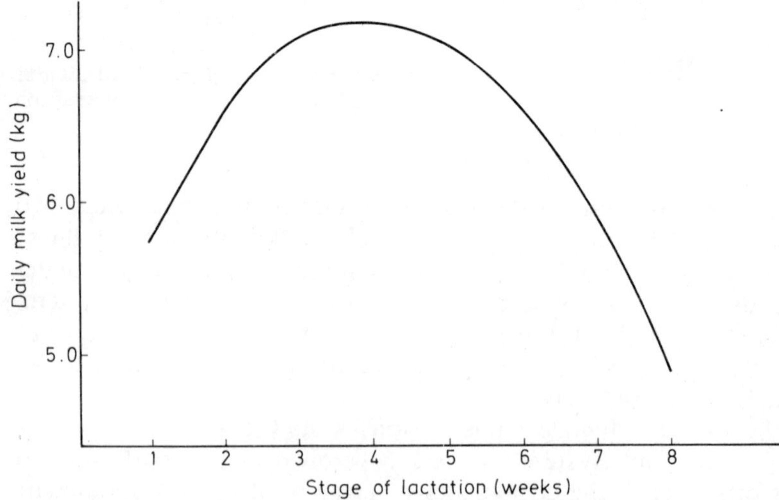

Figure 9.2 The effect of stage of lactation on milk yield (from Elsley, 1970)

At each suckling period the neuro-humoral reflex is initiated by stimuli received by the sensory nerves in the teats or skin of the udder. These carry impulses to the posterior pituitary gland which in turn secretes the hormone oxytocin. This is responsible for the final contraction of the myoepithelial cells surrounding each alveolus thus ejecting milk into the duct system and gland cisterns for withdrawal through the teat orifice by the piglet.

This milk ejection phenomenon is preceded by tactile stimuli from the piglets vigorously 'butting' the sow's udder before finally attaching to the teats and indeed the same butting motion is seen later in each suckling period to ellicit 'stripping' of the udder. On average a sow and litter will suckle between 12 and 18 times in a 24 hour period at regular intervals.

In the first 24 hours or so of lactation the sow secretes colostrum the composition of which is given in *Table 9.1.* The imbibition of

Table 9.1. The composition of colostrum

Component	g/l
Water	698
Lipid	72
Lactose	24
Protein	188
Ash	6

From Long (1961).

sufficient colostrum is of course crucial to piglet survival in the early weeks and this is due to the high content of immunoglobulins present in the protein fraction. The passage of the large molecules through the gut wall intact and into the neonate's blood stream gives it immediate protection from disease challenge. After 24 hours the

Table 9.2. Components of sow's milk

Component	g/l
Water	788
Lipid	96
Lactose	46
Total proteins	61
Calcium	2.1
Phosphorus	1.5
Sulphur	0.8
Vitamin A	1.7 i.u.
Ascorbic acid	0.11
Biotin	14 μg

From Long (1966).

piglets gut is no longer readily permeable to immunoglobulins. If suckling is delayed in the immediate post-partum period this seriously impairs the piglet's chances of survival.

Once lactation is established the composition of milk settles to a lower protein level and lactose levels increase. The composition of sow's milk in mid-lactation is given in *Table 9.2*.

INITIATION OF LACTATION

The change in plasma hormone levels at and around parturition is probably the largest single factor in the triggering of lactation and the change of rapid growth and development of the mammary gland to copious secretion from the germinal epithelial cells within the alveoli.

The change in balance of ovarian steroids probably sets the stage for milk secretion, and the blocking of prolactin inhibiting hormone (PIH) from the hypothalamus allows prolactin and other lactogenic hormones to be released from the anterior pituitary gland. Adrenal steroids are also involved in initiating the process of milk secretion through their effects on general metabolism and on the mammary tissue metabolism.

MAINTENANCE OF LACTATION

As long as the suckling stimulus is maintained then lactation will continue but endocrine factors are also responsible at least in part for the continuation of milk secretion. Most of this endocrine work has been carried out in species other than the pig and there appears to be wide variation between species. However it is known that anterior pituitary hormones and in particular prolactin and growth hormone are necessary to maintain adequate yields, and in addition thyroid hormones and adrenal steroids play a role. With today's systems of earlier weaning, the necessity for having a long and high output lactation is not vital. Indeed in many sows over-production of milk is the major problem rather than the converse.

REGRESSION OF THE MAMMARY GLAND

As soon as the suckling stimulation is removed when the litter is weaned, the alveoli immediately become distended thus compressing the blood capillary system in the process. Within a few days the alveoli cells begin to degenerate and secretory products fill the interstitial spaces and are removed by the lymph system. Phagocytosis then renders the gland down into a non-lobular structure and the parenchyma is reduced to a duct system.

THE OVARIES DURING LACTATION

The corpora lutea of pregnancy are in a state of regression shortly before parturition and by a few days into lactation have become inactive corpora albicantia (Burger, 1952). The ovaries thereafter remain in a dormant state throughout most of lactation unless an exceptionally long lactation length allows the sow to escape from the suppressive action of suckling and show a heat.

POST-PARTUM OESTRUS

It has been observed that a percentage of sows show a behavioural oestrus at or just after farrowing (Warnick, Casida and Grummer, 1950; Burger, 1952; Baker *et al.*, 1953; Self and Grummer, 1958). The actual proportion of sows showing a post-partum oestrus varies considerably between experiments but can be as high as 100 per cent (Burger, 1952). Ovulation seldom occurs during this oestrus period and the overt signs of heat may be due solely to the high levels of circulating oestrogens in the sow's blood stream shortly before parturition. The source of this oestrogen is open to speculation but it is probably placental in origin.

THE GONADOTROPHINS DURING LACTATION

Data relating to the synthesis or release of gonadotrophins are not abundant. Pituitary FSH concentration exhibits change throughout lactation, however, and in fact a high level of activity is seen throughout the suckling period (Lauderdale *et al.*, 1965; Melampy *et al.*, 1966).

A significant drop in luteinizing hormone concentration in the pituitary gland has been noted between the end of pregnancy and day 14 of lactation (Melampy *et al.*, 1966). Pituitary LH also remains very low even after prolonged lactation periods of up to 56 days (Crighton, 1964, 1967). Thus it seems likely that the suckling stimulus inhibits the release of FSH resulting in lack of follicular growth and anoestrus during lactation (Crighton, 1964). Suckling also inhibits LH synthesis in the pituitary, a fact illustrated by recent experiments involving ovariectomy. Indeed when ovariectomy is carried out during the normal oestrous cycle a significant rise in both the synthesis and release of LH is observed (Parlow, Anderson and Melampy, 1964; Rayford, Brinkley and Young, 1971). No such changes occur in sows ovariectomized during lactation (Crighton and Lamming, 1969).

LACTATIONAL ANOESTRUS

In the first week of lactation the number and size of follicles decrease significantly.

After this follicular size increases steadily throughout lactation (Palmer, Teague and Venzke, 1965a) suggesting that there may be a

Table 9.3. The effect of inducing pregnancy in lactation on reproductive performance

Author	*Mode of induction of pregnancy in lactation*	*Treatment*	*Percentage of treated sows showing oestrus in lactation*	*Interval from parturition to oestrus* (days)	*Subsequent litter size*
Burger (1952)	Partial weaning for 12 hours daily between 5 and 31 days post partum		45.5	48.6	
Smith (1961)	Partial separation for 12 hours daily at either 21 days or 31 to 35 days post partum	Separated 21 days	100	30.8	11.2
		Separated 31–35 days	80	39.5	
		Control (unseparated)	–	–	11.4
Crighton (1970)	Partial separation of 12 hours daily plus 1500 i.u. PMSG at 21 days post partum	Treated	77.5	27.5	9.5
		Controls	0	–	9.7
Rowlinson and Bryant (1974)	Grouping sows and litters at 20 days post partum plus introduction of boar 1 day later		100	34.1	11.1

gradual escape from pituitary suppression caused by suckling as lactation proceeds. Some workers have achieved various degrees of success with the induction of oestrus and ovulation in lactation as a means of increasing reproductive efficiency (Crighton, 1970; Rowlinson and Bryant, 1974; Burger, 1952; Smith, 1961). Results from different authors are given in *Table 9.3* which shows that the percentage of sows exhibiting an induced oestrus in lactation varies considerably from 45.5 per cent in Burger's (1952) study to 100 per cent in the work of Rowlinson and Bryant (1974). Differences in the mode of oestrus induction are probably the cause of the variation between experiments. Burger (1952) relied solely on the technique of removing the litters from the dams for 12 hours daily in lactation whereas Rowlinson and Bryant (1974) grouped sows and litters at day 20 of lactation and introduced a boar to the whole group one day later.

Crighton (1970) was unable to achieve a consistent response with either partial weaning (separation for 12 hours daily) or with injection of 1500 i.u. of PMS given at day 21 of lactation. When these treatments were given together however, 77.5 per cent of treated sows showed a fertile heat at an average of 27.5 days after parturition. Litter size was unaffected at the subsequent farrowing. Generally speaking the period of lactation in the sow and gilt is characterized by a suppression of oestrus and ovulation and this is particularly evident in the first 3–4 weeks of lactation. The mechanisms controlling this are not fully understood but certainly the suckling stimulus is a major controlling influence (Peters, First and Casida, 1969). It has been postulated that suckling may inhibit ovulation by depressing the neural stimulus to gonadotrophin synthesis and secretion, but at the same time stimulating prolactin production by depressing the neural inhibition over prolactin secretion (Rothchild, 1967).

9.2 Uterine involution

After parturition and the initiation of suckling, there is a rapid decrease in the length and weight of the uterus (Palmer, Teague and Venzke, 1965a,b). This effect is pronounced in the first week after farrowing and then continues more slowly until a minimum length is reached by day 21–28 of lactation. The effect of stage of lactation on uterine length and weight is shown in *Figure 9.3*.

At parturition and just after the tract is oedematous, a reflection of the high levels of oestrogen to which it has been subjected just prior to this stage. Degeneration of the endometrium is seen during the first week of lactation but by this stage regeneration changes are also evident and complete rebuilding of the endometrium is seen by days 14–21 post partum.

Although complete involution does not occur until about 3 weeks into lactation, it would appear that most of the important changes

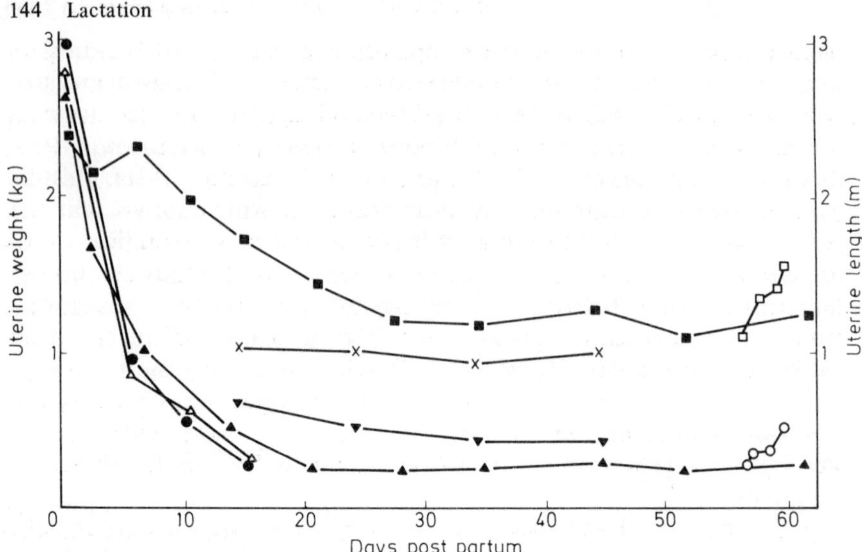

Figure 9.3 Uterine involution. ■——■ Palmer *et al.*, uterine length; △——△ Graves *et al.*, 1967, uterine weight; ▲——▲ Palmer *et al.*, 1965, uterine weight; ●——● Lauderdale *et al.*, 1963, uterine weight; x——x Svajgr *et al.*, 1974, uterine length; ▼—▼ Svajgr *et al.*, 1974, uterine weight; □——□ Palmer *et al.*, 1965, uterine length post weaning (56 days); ○——○ Palmer *et al.*, 1965, uterine weight post weaning (56 days)

have taken place by day 7 of lactation and there seems little reason why the uterus should not function again normally by 2–3 weeks post partum (Palmer, Teague and Venzke, 1965a,b; Graves *et al.*, 1967; Svajgr *et al.*, 1974). The latter point is extremely pertinent to the success or failure of very early weaning systems. If a sow is weaned at day 7 post partum and is mated subsequently at day 15 post partum then fertilized eggs will appear in the uterus at about day 18 post partum. This may therefore be at the limit of uterine capability and embryos may suffer an adverse environment within the lumen of the uterus.

The suckling stimulus is almost certainly involved in the control of uterine involution as suckled sows show a faster rate of involution than sows which have had their litters removed very early in lactation (Graves *et al.*, 1967). This could be solely the result of an absence of ovarian steroid secretion in suckled sows and in fact a quicker regeneration of the uterine gland takes place in sows that have been weaned (Smidt, Thume and Jockle, 1969).

9.3 Factors influencing milk composition and quantity

Although factors such as breed, age and live weight can affect milk quantity and quality, probably the most important factor is nutritional status in terms of energy and protein intake. Energy intake in lactation is variable in practice depending on the size of the suckling litter and an arbitrary 26 MJ digestible energy for the sow plus about 6 MJ per piglet is a frequently used feed scale. Elsley (1970) has highlighted some of the difficulties experienced in estimating precise dietary requirements for lactating sows and gilts and reviewed the subject extensively. It would appear that more than adequate protein intake will be achieved with the above energy scale if the ration contains 150 g/kg of crude protein. Amino acid requirements for lactation have yet to be fully determined although some recent work (Sohail, Lewis and Cole, 1977) has made a superficial investigation into lysine requirements.

9.4 The effects of lactation length on reproductive performance

Although current farm practice includes the complete range of systems from 7-day weaning to 56-day weaning, there are probably three basic systems of production to consider. First, the traditional 5–8 week weaning system, second, the 3-week weaning system now becoming very popular, and also the very early weaning system involving removal of piglets from the dam at between 7 and 14 days after parturition.

The advantages with shorter lactations are associated with the much reduced average farrowing interval giving potentially a considerable boost to the numbers of litters produced per sow per year and hence the raising of annual sow productivity in terms of piglets born alive per sow per year. In reality, however, limitations imposed by the sow's physiological mechanisms may restrict the realization of this potential.

ENDOCRINE RELATIONSHIPS

Very few studies have included an investigation into the endocrine relationships of sows subjected to lactation lengths of less than 8 weeks. It is to be hoped that with the advent of widespread use of radioimmunoassay techniques (Rayford, Brinkley and Young, 1971; Niswender, Reichert and Zimmerman, 1970; Haresign *et al.*, 1975) this situation will be rectified. Residual pituitary concentration of the gonadotrophins after lactation lengths of 1, 6 or 11 days however has been studied by Lauderdale *et al.* (1965). In these studies sows were slaughtered 5 days after weaning and were compared with groups of sows slaughtered at the same time post partum but which were suckled through until slaughter. No significant effects of weaning or suckling on pituitary LH activity were detected.

This experiment does not therefore provide comparison of the different postweaning response for gonadotrophin activity between a short and a traditional weaning system but a tentative comparison between this work and the work of Crighton (1967) has been suggested by Polge (1972). He concluded that the concentration of LH in the pituitary increases very much less in sows that were weaned within 1–11 days post partum than in sows weaned at 56 days post partum. Also FSH concentration in the pituitary dropped in sows weaned shortly after parturition but remained constant in sows weaned at 8 weeks. Follicular development during the 5 days post weaning appears to be very much less in sows that were weaned after a short lactation period. All of this suggests that the very early weaned sow shows a smaller postweaning pituitary response than a sow subjected to a more traditional weaning age.

In an attempt to manipulate the sow's endocrine system by exogenous application, Peters *et al.* (1969) removed piglets at birth and gave a daily intramuscular injection of FSH for 3 days immediately post partum followed by an injection of bovine pituitary extract on day 5 post partum. This group of sows was then compared with a similar group which was suckled until slaughter at day 8 post partum. All sows were inseminated on days 3, 4, 5 and 6 post partum to ensure fresh semen was in the tract at the time of ovulation. Results showed that in both groups ovulation occurred at approximately 40 hours after the first injection, but in neither group were fertilized ova observed. Suckled sows however had a greater number of ovulations. Half of the sows weaned at birth also had cystic ovaries and/or remained in constant oestrus.

Kirkpatrick *et al.* (1965) in a similar experiment to that of Lauderdale *et al.* (1965) also gave injections of pituitary extract for 5 days prior to slaughter at 6, 11 or 16 days post partum and compared suckling to slaughter with weaning 5 days before slaughter. The response in terms of follicular growth was much higher in sows weaned at 1 day post partum than in sows weaned at 11 days post partum. Ovulation response was also greater for the sows injected immediately after parturition.

It may be concluded from these experiments that very short lactations are associated with serious postweaning endocrine disturbances which cannot be rectified by a simple hormonal application. A detailed and precise plasma hormonal profile of the early weaned sow is needed before we can begin to formulate remedial measures. As yet such a profile is not available.

Moody and Speer (1971) have looked at luteal progesterone content at slaughter 25 days post coitum after lactation lengths of either 14, 21 or 28 days and found no significant differences. Luteal content however does not give an indication of plasma level which is the real indicator of endocrine activity. Varley (1976) has observed that differences may exist in the response of plasma progesterone to

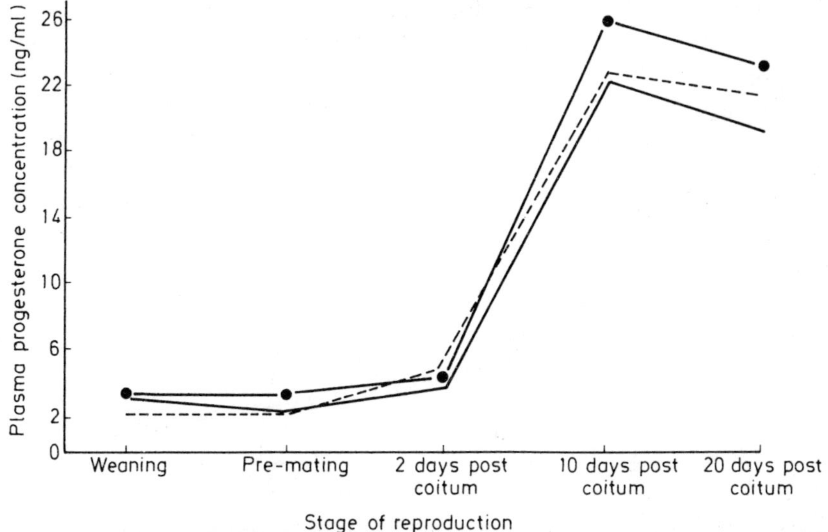

Figure 9.4 The effect of lactation length on plasma progesterone concentration. —— 7-day weaned; ● 21-day weaned; – – – – 42-day weaned

different lactation lengths and this is illustrated in *Figure 9.4*. The 42-day weaned group follows very closely the 7-day weaned group until 10 days post coitum but then plasma progesterone for the 7-day weaned sows drops away much faster than for the 42-day group up to day 20 post coitum. The 21-day group is overall much higher in plasma progesterone than either of the other two groups. This was probably a result of the much higher ovulation rate for this group due to chance.

ENDOMETRIAL REGENERATION AND UTERINE INVOLUTION

The uterus with a conventional 8-week weaning system rapidly loses length and weight in the first 2–3 weeks post partum and then remains at a constant value until weaning after which increased weight and length occur in response to circulating oestrogens.

Graves *et al.* (1967) have compared the effect of weaning after a short lactation (0, 6 and 11 days) with suckling to the same stage at slaughter which was 5 days following weaning for each treatment group. They reported that weaning at any of these times resulted in a slowing down of uterine weight loss (*Figure 9.5*). The differences were not great however and in one comparison loss in weight was greater for the weaned sows than for the sows suckled over the same period. Svajgr *et al.* (1974) have observed also (*Figure 9.3*) that sows which were weaned following a 2-day lactation length (Group A) had significantly greater uterine weights than the other three groups at the

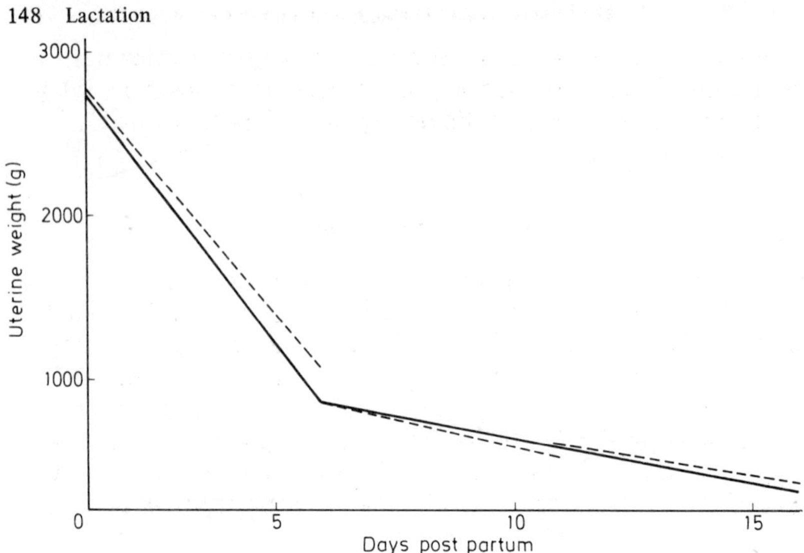

Figure 9.5 The effect of suckling or piglet removal on uterine weight loss after parturition (from Graves *et al.*, 1967). ——— suckled until slaughter; – – – – weaned 5 days before slaughter

same stage relative to mating. It seems likely therefore that very short lactations slow down the rate of loss in weight of the uterus after parturition and if mating occurs very close to farrowing the developing embryos may leave the oviducts at a stage when the uterus is not ideally conditioned for optimum embryonic development.

Despite this conclusion glandular regeneration of the endometrium begins by day 7 post partum and is apparently complete by day 21 post partum (Palmer, Teague and Venzke, 1965b). Graves *et al.* (1967) observed also that from day 6 post partum onwards loss in weight of tissues is mostly at the expense of the myometrium while the endometrium is largely unaffected. Smidt, Thume and Jochle (1969) have found that sows weaned at 4 days compared with suckled sows showed an increased area of uterine glands as a percentage of the total area of the endometrium. Endometrial repair therefore may proceed at a faster rate following weaning in early lactation. Svajgr *et al.* (1974) have added to this finding with the observation that sows weaned after a 2-day lactation have a fully regenerated endometrium at day 15 post partum. Also because uterine gland diameters and heights were unaffected by the length of the suckling period normal uterine gland function appeared to be restored by the time fertilized ova reached the uterus in sows bred on the first post-weaning oestrus following a 2-day lactation length.

However, a difference in oviducal epithelial height was evident at day 15 post partum between the 2-day weaned group and the rest (13-, 14- and 35-day lactation periods). The authors in this study have concluded therefore that the uterus is only structurally ready to accept blastocysts when the sows are bred on the first postweaning oestrus following a 13-day lactation period. For shorter lactations incomplete uterine involution and lack of tissue repair will be responsible for increased embryonic loss.

THE INTERVAL FROM WEANING TO OESTRUS

The interval from weaning to oestrus has been shown in a number of studies to be largely a function of the preceding lactation length (Aumaitre, 1972; te Brake, 1972; Dyrendahl *et al.*, 1958; Moody and Speer, 1971; Self and Grummer, 1958; Van der Heyde, 1972; Svajgr *et al.*, 1974; Cole, Varley and Hughes, 1975; Varley and Cole, 1976a). An illustration of the relationship is given in *Figure 9.6* showing quite clearly that as lactation length is reduced the interval from weaning to the onset of heat increases (*see also Table 9.4*).

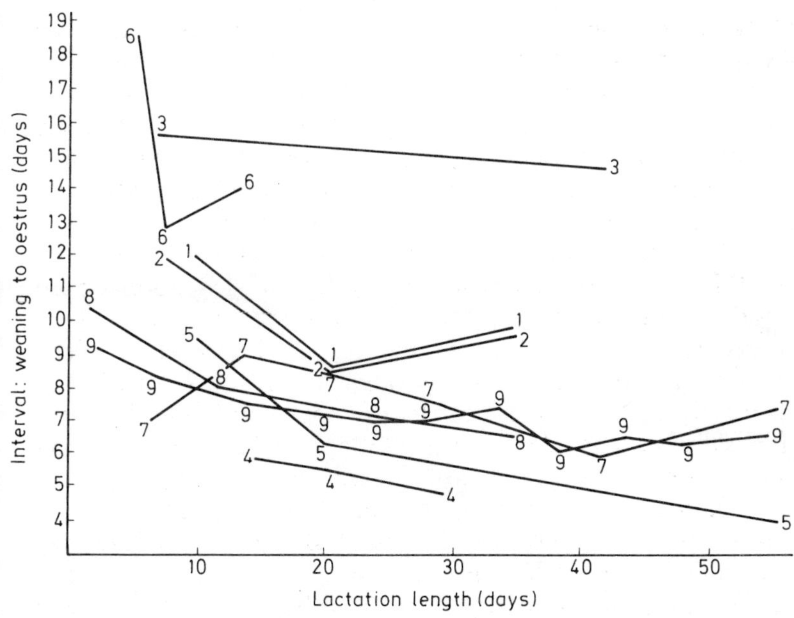

Figure 9.6 The effect of lactation length on the interval from weaning to oestrus. (Numbers refer to references in *Table 9.4*)

A view held by pig farmers has been that the sow could never show a heat prior to 3 weeks post partum. This view is discounted by the data presented in *Figure 9.6* and it would appear that even for lactation lengths of around 10 days the sow will show a heat at about 9–10 days post weaning. The precise nature of the relationship between lactation length and the weaning to oestrus interval has

Table 9.4. The effect of lactation length on the interval from weaning to the next oestrus

Author and reference (Figure 9.6)	*Age at weaning* (days)	*Parity number*	*Weaning to oestrus interval* (days)
Aumaitre (1972)	7–10		11.7
(1)	21		8.5
	35		9.8
Aumaitre and	10	0	10.6
Rettigliati (1972)	10	1	11.4*
(2)	10	2	11.4
	10	3,4	13.1
	21	0	9.5
	21	1	8.1
	21	2,3	9.0
	35	1,2,3	9.3
te Brake (1972)	7	n	15.3
(3)	7	n + 1	15.2
	7	n + 2	16.0
	42	n	15.0
	42	n + 1	15.7
	42	n + 2	12.9
Dyrendahl *et al.* (1958)	4.1		10.6
Moody and	14		5.9
Speer (1971)	21		5.6
(4)	28		4.8
Self and Grummer	10		9.4
(1958)	21		6.2
(5)	56		4.0
Van der Heyde	4.2		18.6
(1972)	8.0		12.9
(6)	13.0		14.1
Banjac *et al.*	19.9		34.1*
(1968)	29.9		35.5
Lynch (1965)	0–19		12.4
	31–38		5.4
Puyaoan and Castillo	7		7.0
(1963)	14		9.0
(7)	21		8.5
	28		7.6
	35		10.4

	42	6.0
	56	7.25
Svajgr *et al.* (1974) (8)	2	10.1
	13	8.2
	24	7.1
	35	6.8
Smidt *et al.* (1965) (9)	0–5	9.06
	6–10	8.03
	11–15	7.79
	16–20	7.43
	21–25	7.25
	26–30	7.00
	31–35	7.44
	36–40	6.42
	41–45	6.50
	46–50	6.43
	51–55	6.53

* Estimate of weaning to oestrus interval from given farrowing interval.

been evaluated by Cole, Varley and Hughes (1975) using a regression approach and it has been determined that:

$$\text{Log } Y = 0.931 - 0.0077\, X$$

where Y = the interval from weaning to oestrus and
X = lactation length.

This is illustrated in *Figure 9.7*.

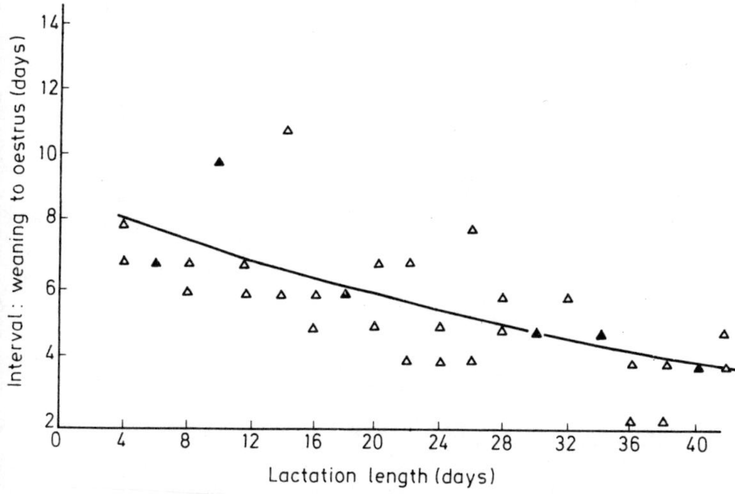

Figure 9.7 The effect of lactation length on the weaning to remating interval. △ one animal; ▲ two animals

CONCEPTION RATE

Reported data concerning the effect of lactation length on the conception rate are given in *Table 9.5* and illustrated in *Figure 9.8.*

As lactation length is reduced, conception generally seems to be adversely affected although there is some doubt as to the magnitude of the effect. Aumaitre (1972) and Aumaitre and Rettigliati (1972) have observed that a drop of approximately 10 per cent in conception is associated with a change in lactation length from 35 to 10 days whereas te Brake (1972) has observed a somewhat larger drop over a similar range in lactation length. Van der Heyde (1972) in a field survey on some 284 sows weaned between 0 and 20 days post partum has observed that only 5.6 per cent of sows were culled because they were non-pregnant 75 days after farrowing. He goes on to conclude that there is considerable variation between farms in this respect and it should be possible to achieve as good conception rates from very early weaning as from the more traditional systems. He adds that the low conception rates on some farms were associated

Table 9.5. The effect of lactation length on conception rate

Author and reference (*Figure 9.8*)	*Age at weaning* (days)	*Parity number*	*Conception rate* (%)
Aumaitre (1972)	7–10		86.0
(1)	21		95.0
	35		97.6
Aumaitre and	10		77.3
Rettigliati (1972)	21		85.4
(2)	35		87.8
te Brake (1972)	7	n	60.6
(3)	7	n + 1	42.9
	7	n + 2	56.5
	42	n	73.3
	42	n + 1	81.6
	42	n + 2	82.9
Dyrendahl (1958)	4.1		81.0
Moody and	14		66.7
Speer (1971)	21		83.3
(4)	28		83.3
Svajgr *et al.* (1974)	2		68*
(5)	13		92*
	24		100*
	35		100*

* Data based on non-return rate.

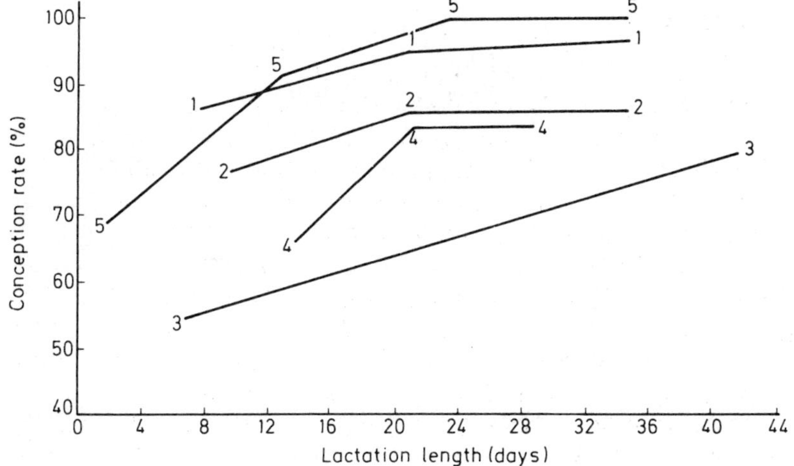

Figure 9.8 The effect of lactation length on conception rate. (Numbers refer to references in *Table 9.5*)

with incorrect frequency and timing of inseminations. Cole, Hughes and Varley (1975) and Varley and Cole (1976a) have also observed no reduction in farrowing rates or conception rates with decreased lactation length.

FERTILIZATION RATE

Fertilization rate and its relationship to the length of the previous lactation has been examined by a number of workers. Most authors conclude that there is no relationship between the two. Baker *et al.* (1953) found that the percentage of ova fertilized 24 hours after ovulation was similar for sows weaned after birth or 2-day lactation periods and Self and Grummer (1958) have observed that following 10-, 21- and 56-day lactations, fertilization rates were 93.4 per cent, 90.6 per cent or 98.1 per cent respectively and the results were not significant. For lactation lengths of 2, 13, 24 and 35 days Svajgr *et al.* (1974) obtained fertilization rates of 81.9 per cent, 86.3 per cent, 96.5 per cent and 98 per cent respectively but attribute this result to two sows in the 2-day group and one in the 13-day group which showed a complete absence of cleaved ova at day 3 post coitum.

LACTATION LENGTH AND PROLIFICACY

As lactation length is reduced most authors agree there is a concomitant reduction in the number of piglets born per litter in the next parity (Aumaitre, 1972; te Brake, 1972; Moody and Speer, 1971; Smidt, Scheven and Steinbach, 1965; Cole, Varley and

Hughes, 1975; Varley and Cole, 1976a). The size of this reduction in prolificacy however shows some disparity between experiments which could reflect possible breed and environmental differences.

These experiments are summarized in *Table 9.6* and illustrated in *Figure 9.9.* It seems that over the range from 6-week lactation length to 3-week lactation length there is little if any change in litter size but below this length litter size drops off considerably. Varley (1979) has concluded that the 3-week weaned sow probably incurs a reduction of up to 0.2 piglets per litter in prolificacy and Varley and Cole (1976a,b) have shown that a sow weaned at 7–10 days post partum will show a decrease of between 1.5 and 2.0 piglets per litter at the next farrowing. Despite these reports there have been commercial reports (Pay, 1973; Looker, 1974) suggesting that very early weaning

Table 9.6. The effect of lactation length on the subsequent litter size

Author and reference (*Figure 9.9*)	*Age at weaning* (days)	*Litter size*
Aumaitre (1972)	7–10	9.3
(1)	21	10.4
	35	10.0
Aumaitre and Rettigliati (1972)	10	9.4
(2)	21	10.4
	35	11.5
Moody and Speer (1971)	14	9.3
(3)	21	10.7
	28	12.2
Van der Heyde (1972)	4.2	8.7
(4)	8.0	8.7
	13.0	9.9
Smidt *et al.* (1965)	0–5	7.3
(5)	6–10	8.1
	11–15	9.0
	16–20	8.8
	21–25	9.7
	26–30	9.7
	31–35	10.0
	36–40	9.9
	41–45	10.1
	46–50	10.0
	51–55	9.6
te Brake (1972)	7	9.0
(6)	42	11.1

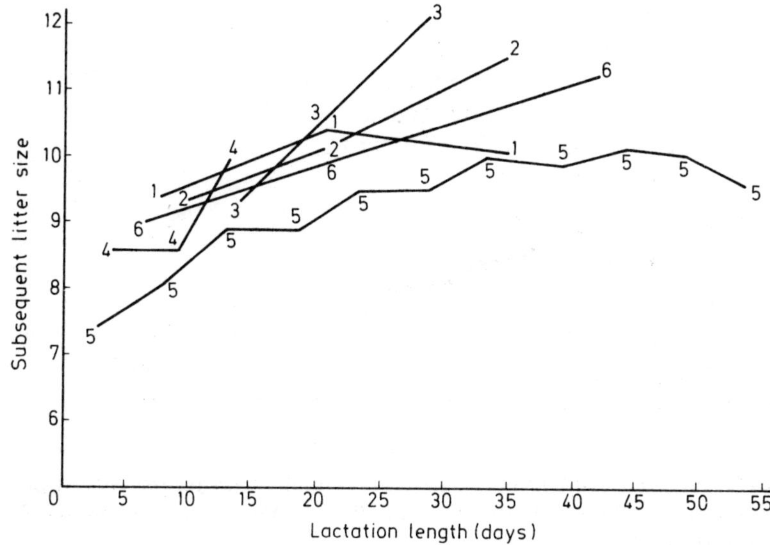

Figure 9.9 The effect of lactation length on subsequent litter size. (Numbers refer to references in *Table 9.6*)

systems can work with no apparent adverse effect on prolificacy. Some of this discrepancy may be accounted for by differences in feeding systems but Varley and Cole (1976a) have observed that varying the feed plane from 2 to 4 kg/day in lactation and from weaning to service has no effect on prolificacy for the early weaned sow. In contrast however Hughes (unpublished data) has observed that when a much higher feed plane of 8 kg/day was employed in the service period then there was no reduction in litter size.

Varley and Cole (1976b) have shown that ovulation rate is unaffected by lactation length but the ability of embryos to survive the first 3 weeks of gestation is very severely impaired by subjecting sows to a very short lactation length. In fact embryo survival dropped from 80 per cent to 60 per cent as lactation length changed from 6 weeks to 7 days. Clearly there is much that remains to be understood in this area but it does seem that a solution to the problem of litter size reduction for the early weaned sow might be forthcoming in the future.

9.5 Conclusions

Figure 9.10 shows the likely effect of a range in lactation length on annual sow productivity. It can be seen that sow productivity in terms of piglets produced per sow per year is maximized for lactation lengths of 3–4 weeks and for shorter lactation lengths annual sow productivity falls off because of the significant reduction in litter size.

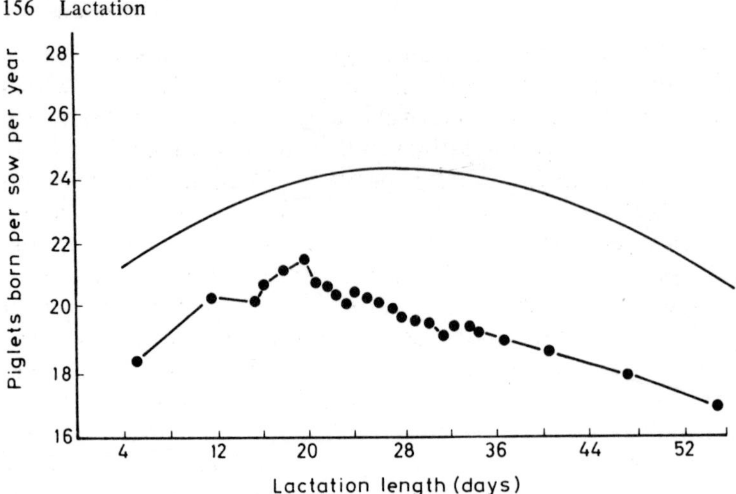

Figure 9.10 The effect of lactation length on annual sow productivity. Upper curve, from Smidt, Scheven and Steinbach, 1965; lower curve, from Aumaitre, Perez and Chauvel, 1975

The results of a series of experiments carried out at Nottingham University looking into this problem are given in *Figure 9.11*. From this work it transpires that the non-significant difference between the early weaned sow and the 42-day weaned sow for numbers of viable embryos at 9 days post coitum (Varley and Cole, 1978) is in striking contrast to the large significant difference at 20 days post

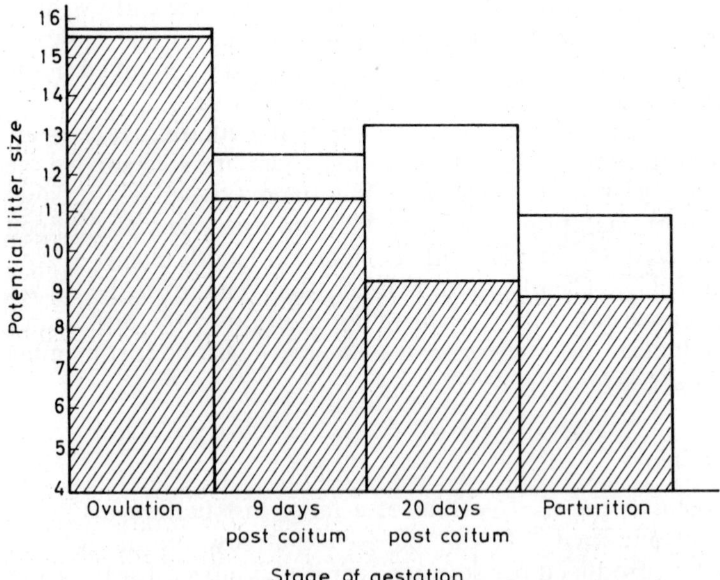

Figure 9.11 Litter size throughout gestation for the early weaned sow. Shaded, 7–10 day weaned; open, 42-day weaned

coitum between the two lactation length systems. This difference is then reflected in numbers born alive at parturition. In other words pre-implantation losses are relatively unimportant for the early weaned sow compared to losses incurred at or around implantation.

At the present time therefore a producer looking for maximum productivity from his sows should be operating a 3–4 week weaning age. Above this level the number of litters per sow per year falls off and below this litter size drops off for the reasons previously mentioned. Despite this conclusion there may still be good reasons in a practical context why it may be better for some producers to adopt a more conventional weaning age such as 5 weeks. Available housing might not be adequate for 3-week weaned piglets without additional investment, and available labour might not be sufficiently attuned to handling and rearing them. Certainly if these two aspects are not right then a 3-week weaning system can fall down because of high mortality of piglets and inadequate growth performance.

To conclude this chapter, therefore, it is evident that there is probably more scope for achieving higher levels of sow output by varying the lactation length than with any other form of reproductive control available but there are still many major obstacles to the exploitation of the sow's full potential. When these are finally overcome with increased research effort, three litters per sow per year and 28 or 29 piglets per sow per year will be the normal annual sow productivity.

9.6 References

AUMAITRE, A. (1972). *23rd Annual Meeting. E.A.A.P. Commission in Pig Production (Verona)*

AUMAITRE, A., PEREZ, J.M. and CHAUVEL, J. (1975). *Journées de la recherche porcine en France*, pp. 52–67, Paris, L'Institat Technique du Porc

AUMAITRE, A. and RETTIGLIATI, J. (1972). *Annls Zootech.* **21**, 634–635

BANJAC, J., BANJAC, D. and SVIBEN, M. (1968). *Anim. Breed. Abstr.* **36**, 3857

BAKER, L.N., WOEHLING, H.L., CASIDA, L.E. and GRUMMER, R.H. (1953). *J. Anim. Sci.* **12**, 33–38

te BRAKE, J.A.A. (1972). *23rd Annual Meeting. E.A.A.P. Commission in Pig Production (Verona)*

BURGER, J.F. (1952). *Onderstepoort. J. vet. Res. Suppl.* **2**, 3–218

COLE, D.J.A., VARLEY, M.A. and HUGHES, P.E. (1975). *Anim. Prod.* **20**, 401–406

CRIGHTON, D.B. (1964). *Proc. 5th int. Congr. Anim. Reprod. A.I. Trento.* **2**, 349–354

CRIGHTON, D.B. (1967). In *Reproduction in the Female Mammal* (Ed. by G.E. Lamming and E.C. Amoroso), pp. 223–238. London, Butterworths

CRIGHTON, D.B. (1970). *J. Reprod. Fert.* **22**, 223–231

CRIGHTON, D.B. and LAMMING, G.E. (1969). *J. Endocr.* **43**, 507

CROSS, B.A., GOODWIN, R.F.N. and SILVER, I.A. (1958). *J. Endocr.* **17**, 63–74

DYRENDAHL, S., OLSSON, B., BJORCK, G. and EHLERS, T. (1958). *Acta Agric. scand.* **8**, 3–19

ELSLEY, F.W.H. (1970). In *Lactation* (Ed. by I.R. Falconer.), pp. 393–491. London, Butterworths

FRANDSON, R.D. (1974). *Anatomy of Farm Animals*, 2nd edn. Philadelphia, Lea and Febiger

GRAVES, W.E., LAUDERDALE, J.W., KIRKPATRICK, R.L., FIRST, N.L. and CASIDA, L.E. (1967). *J. Anim. Sci.* **26**, 365–369

HARESIGN, W., FOSTER, J.P., HAYNES, N.B., CRIGHTON, D.B. and LAMMING, G.E. (1975). *J. Reprod. Fert.* **43**, 269–279

KIRKPATRICK, R.L., LAUDERDALE, J.N., FIRST, N.L., HAUSER, E.R. and CASIDA, L.E. (1965). *J. Anim. Sci.* **24**, 1104–1106

KUHN, N.J. (1970). In *Lactation* (Ed. by I.R. Falconer.), pp. 161–176. London, Butterworths

LAUDERDALE, J.W., KIRKPATRICK, R.L., FIRST, N.L., HAUSER, E.R. and CASIDA, L.E. (1965). *J. Anim. Sci.* **24**, 1100–1103

LONG, C. (1961). *Biochemist's Handbook.* London, E. and F.N. Spon

LONG, C. (1966). *Biochemist's Handbook.* London, E. and F.N. Spon

LOOKER, M. (1974). *Pig Fmg* **22(10)**, 66–67

LYNCH, G. (1965). *Meld. Norg. LandbrHøisk* **44**, No 4

MELAMPY, R.M., HENDRICKS, D.M., ANDERSON, L.L., CHEN, C.L. and SCHULTZ, J.R. (1966). *Endocrinology* **78**, 801–804

MOODY, N.W. and SPEER, V.C. (1971). *J. Anim. Sci.* **32**, 510–514

NISWENDER, G.D., REICHERT, L.E.J.R. and ZIMMERMAN, P.R. (1970). *Endocrinology* **87**, 576–580

PALMER, W.M., TEAGUE, H.S. and VENZKE, W.G. (1965a). *J. Anim. Sci.* **24**, 541–545

PALMER, W.M., TEAGUE, H.S. and VENZKE, W.G. (1965b). *J. Anim. Sci.* **24**, 1117–1125

PARLOW, A.F., ANDERSON, L.L. and MELAMPY, R.M. (1964). *Endocrinology* **75**, 365–376

PAY, M.G. (1973). *Vet. Rec.* **92**, 255–259

PETERS, J.B., FIRST, N.L. and CASIDA, L.E. (1969). *J. Anim. Sci.* **28**, 537–541

POLGE, C. (1972). *Proc. Br. Soc. Anim. Prod.* **1**, 5–18

PUYAOAN, R.B. and CASTILLO, L.S. (1963). *Philipp. Agric.* **47**, 32–44

RAYFORD, P.L., BRINKLEY, H.J. and YOUNG, E.P. (1971). *Endocrinology* **88**, 707–713

ROTHCHILD, I. (1967). In *Reproduction in the Female Mammal* (Ed. by G.E. Lamming and E.C. Amoroso.), pp. 30–54. London, Butterworths

ROWLINSON, P. and BRYANT, M.J. (1974). *Proc. Br. Soc. Anim. Prod.* **3**, 93

SELF, H.L. and GRUMMER, R.H. (1958). *J. Anim. Sci.* **17**, 862–868

SMIDT, D., SCHEVEN, B. and STEINBACH, J. (1965). *Züchtungskunde* **37**, 23–36

SMIDT, D., THUME, O. and JOCHLE, W. (1969). *Züchtungskunde* **41**, 36–45

SMITH, D.M. (1961). *N. Z. J. agric. Res.* **4**, 232–245

SOHAIL, M.A., LEWIS, D. and COLE, D.J.A. (1977). *Br. J. Nutr.* **39**, 463

SVAJGR, A.J., HAYS, V.W., CROMWELL, G.L. and DUTT, R.H. (1974). *J. Anim. Sci.* **38**, 100–105

VAN DER HEYDE, H. (1972). *Proc. Br. Soc. Anim. Prod.* **1**, 33–36

VARLEY, M.A. (1976). PhD Thesis. University of Nottingham

VARLEY, M.A. (1979). *Pig Fmg* **27(2)**, 34–35

VARLEY, M.A. and COLE, D.J.A. (1976a). *Anim. Prod.* **22**, 71–77

VARLEY, M.A. and COLE, D.J.A. (1976b). *Anim. Prod.* **22**, 79–85

VARLEY, M.A. and COLE, D.J.A. (1978). *Anim. Prod.* **27**, 209–214

WARNICK, A.C., CASIDA, L.E. and GRUMMER, R.H. (1950). *J. Anim. Sci.* **9**, 66

Chapter 10

The weaning to conception interval

The mean weaning to conception interval is determined by three components. First, the average interval from weaning to the appearance of heat, second, the proportion of sows which return to heat 3 weeks later and are then mated again successfully and third, the sows and gilts which following weaning take a vastly extended time to return to effective service following weaning. Some in fact may be completely anoestrous and will be culled from the herd and replaced. In this last category may be placed sows which repeatedly fail to hold to service and in addition any sows which are found non-pregnant at or near the next expected farrowing date.

If it were not for sows in the second and third categories, the average interval from weaning to conception would be of the order of 4–5 days and the average sow would spend little time being non-productive. In reality this is not the case and on many pig farms the number of these so-called 'empty days' may be very large indeed.

For example, consider the typical weaner producer weaning at 5 weeks. With a gestation length of 115 days and a target weaning to conception interval of 5 days, sows farrow every 155 days or each sow has 2.35 litters every year. National estimates of farrowing index, however, show that on average producers only achieve 2.0 litters per sow per year. Part of this discrepancy is accounted for by the normal biological limitations imposed by the physiology of the sow but the larger part is due to poor management.

Varley (1979) has estimated that an increase of 15 per cent in anoestrous sows reduced sow productivity by 5.7 per cent and an increase of 15 per cent in sows not effectively served and found non-pregnant at term reduces sow productivity by 9.6 per cent. These two factors are probably largely under management control.

The present chapter will therefore review factors affecting the components of the interval from weaning to conception and will attempt to provide some practical recommendations for keeping the number of these 'empty days' as small as possible.

10.1 Physiology in the weaning to remating interval

Many important aspects of this interval relate to ovulation and fertilization which have been dealt with in Chapters 5 and 6. There is a tremendous burst of endocrine and physiological activity in this interval, despite its short duration compared to the pig's reproductive life cycle.

In the immediate postweaning hours the first phenomenon to consider is the cessation of milk production. The alveoli become very distended causing the udder to appear hard and swollen for the first 24 hours after weaning. The lactation process gradually shuts down following this because milk secretion is in equilibrium with the reabsorbtion into the blood stream of secretory products. By 2–3 days after weaning, the udder is less distended as the alveolar tissue begins to break down.

There has been some work done on the practical consequences of restricting or not the feed and/or water to the weaned sow for 24 hours following weaning. The idea is to encourage a rapid and stress-free end to lactation with a view to enhancing the subsequent post-weaning ovarian response. Brooks and Cole (1972) have shown no improvement in reproductive performance (including the length of the remating interval) by withholding feed for 24 hours post weaning. In addition it is now established that there is no benefit gained from withholding water from the sow in the immediate postweaning period. Previously it had been thought that such a practice would help the udder regress and assist the speedy return to oestrus by the sow but this is not so.

ENDOCRINE ACTIVITY

Pituitary FSH levels rise very little in the immediate postweaning period but pituitary LH rises significantly (Crighton, 1967). Increase in the production and storage of LH must therefore be a feature of the sow's postweaning endocrine activity as no peripheral LH activity has been observed until immediately before ovulation (Parlow, Anderson and Melampy, 1964). However FSH, which does not increase its concentration in the pituitary gland from weaning onwards, must be undergoing rapid production and release from the pituitary from weaning onwards in view of the fact that the number and size of the follicles on the ovaries increase in the same period. This may be due in part to the changing sensitivity of the ovaries to FSH throughout the same period.

Oestrus occurs at the culmination of the rise in circulating oestrogens secreted from the ovarian follicles as a result of FSH activity. At and just prior to oestrus pituitary levels of both FSH and LH fall off quickly (Parlow, Anderson and Melampy, 1964; Polge, 1972).

As yet there is little evidence showing the endocrine activity in sows that fail to hold to service at the postweaning oestrus or for sows showing the various forms of infertility such as silent heats, anoestrus and so on. At first sight it would seem probable that endocrine imbalance plays a role in at least a proportion of cases of infertility of this kind.

UTERINE CHANGES

It has been shown by Palmer, Teague and Venzke (1965) that following a 56-day lactation period the uterus begins to gain weight and length in the first few days after weaning probably as a result of gradually increasing levels of oestrogens in the circulation. The response of the uterine tissues to a much reduced lactation period has yet to be determined, and is becoming an important question with the now widespread adoption of 3-week weaning systems.

10.2 Factors influencing the weaning to oestrus interval

The interval from weaning to oestrus is influenced by a number of variable factors and not the least of these is the length of the preceding lactation (Chapter 9). In addition the *apparent* weaning to oestrus interval often depends on the ability of stockmen to detect heat and also, in practice, upon the efficiency of the recording scheme.

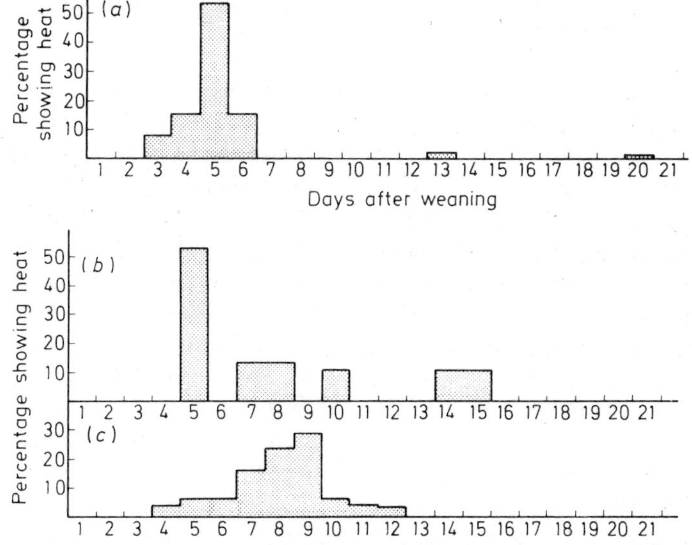

Figure 10.1 The distribution of the time of heat onset. (a) 42-day weaning; (b) 21-day weaning; (c) 7–10 day weaning

The interval from weaning to oestrus has however a modal value of 4–5 days for lactations of 5–8 weeks but there is a wide spread around these modal values and some sows normally show a heat on day 2 following weaning and some on day 18 following weaning. When heat is first observed at (say) day 23 following weaning this begins to look more like a heat missed the first time.

Figure 10.1, from the University of Nottingham data, illustrates the distribution of heat appearance for three different lactation lengths. Clearly a great deal of spread exists with every system.

EFFECTS OF GENOTYPE

Genotype affects the length of the weaning to oestrus interval and breed differences have been noted in two studies. *Table 10.1* gives the results from these experiments and indicates that the Large White breed of pig appears to return to heat quicker than the Large Black breed and this difference is quite substantial (Burger, 1952).

Table 10.1. Breed differences for the interval from weaning to the onset of oestrus

Breed	*Interval* (days)	
	Burger (1952)	*Dyck (1971)*
Large White	7.85	
Large Black	16.08	
Yorkshire		5.5
Lacombe		14.3

Dyck (1971) reports that Yorkshire sows return to oestrus faster than the Lacombe breed and again the difference was large. Clark *et al.* (1972) found that the Yorkshire breed had a mean interval of 5.1 days which is similar to Dyck's (1971) estimate. In the report of Clark *et al.* (1972) however no difference was reported between the performance of the Yorkshire breed and the Poland China breed and in addition the Poland China × Yorkshire crossbred sow performed exactly as the parent breeds and therefore little evidence of a heterotic effect was seen.

Differences between the Yorkshire, the Large White and the various Landrace breeds may be significant but in practice the differences are very small.

THE EFFECTS OF AGE

The effects of age in the multiparous sow are very closely related to the number of parities and in the experiments reported to date no attempt has been made to separate out these two components.

From the literature it would seem that as parity number increases there is a more consistent return to oestrus by the sow (i.e. less spread like that in *Figure 10.1*). Du Mesnil du Buisson and Signoret (1968) have observed that for gilts weaned after their first litter, 25.4 per cent came back into oestrus within 9 days of weaning whereas with sows weaned after six or more litters, 55.3 per cent came back into heat within 9 days. This is in general agreement with Aumaitre, Perez and Chauvel (1975) who showed that as parity number increased from 1 to 6+ the interval from weaning to conception decreased from 28.3 days to 17.6 days. The latter report however includes conception rate factors and as a consequence is not the simple reflection of parity on the weaning of oestrus interval.

THE EFFECTS OF SEASON

The effects of season on the weaning to oestrus interval are in part an effect of temperature and also an effect of changing light patterns. It is important to note at this point, however, that effects due to seasonality may be less now than they were previously because of the widespread use of highly intensive housing systems, including controlled temperature and in many cases windowless housing. As yet there is little evidence to suggest what the correct light pattern, light intensity or temperature may be to achieve the shortest and most consistent return to oestrus.

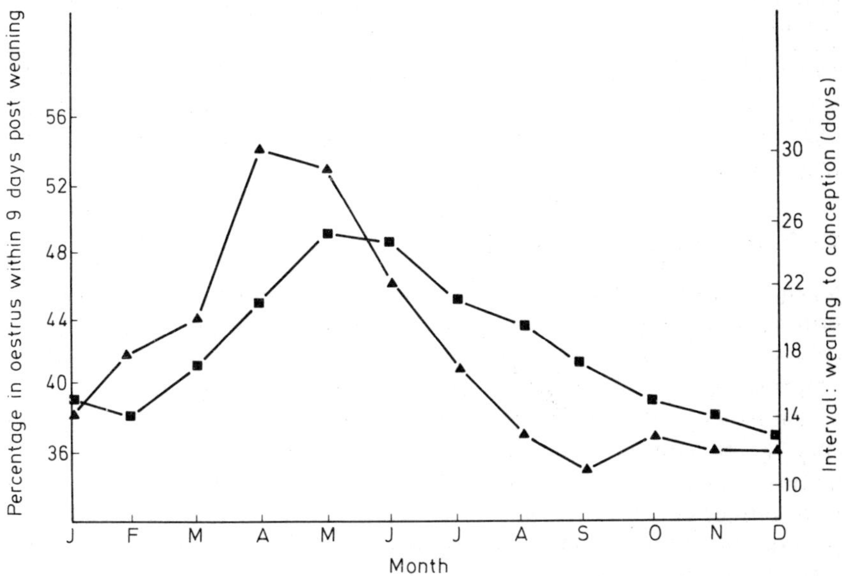

Figure 10.2 The effect of month of year on the postweaning return to oestrus. ■ from Legault, Dagorn and Tastu, 1975; ▲ from Du Mesnil du Buisson and Signoret, 1968

The colder months of the year tend to invoke an inconsistent response measured by the percentage of sows returning to heat within a given time (Du Mesnil du Buisson and Signoret, 1968), but overall the mean value is lower for the winter months than the summer (Legault, Dagorn and Tastu, 1975). This is illustrated in *Figure 10.2.*

EXOGENOUS HORMONE TREATMENT

The use of exogenous hormones as a means of shortening this interval has been examined by Longnecker and Day (1958). An injection of 1200 i.u. of PMS on the day of weaning after a lactation period of 4 or 8 weeks caused a significant reduction in the average time to return to oestrus from 5.4 days in untreated animals to 4.8 days in treated ones. It would appear, however, that the technique, although consistent, is probably not cost effective or even of much practical value.

SCALES OF FEEDING

A significant effect of scale of feeding from weaning to remating on the interval from weaning to conception was found by Brooks and Cole (1972), who reported a drop from 21.6 ± 3.0 days for sows fed 1.8 kg/day to 9.2 ± 2.2 days for primiparous sows fed 3.6 kg/day between weaning and remating. In a later report however Brooks *et al.* (1975) were unable to repeat this observation in a large scale coordinated field study using multiparous sows. The latter observation is in agreement with the work of Clark *et al.* (1972) and Dyck (1972) who also could find no effect of postweaning feed level on the interval from weaning to oestrus.

10.3 Practical recommendations

Minimization of the interval from weaning to conception is of paramount importance if maximization of sow productivity is to be achieved. It is obvious from national statistics that a vast number of producers are not succeeding in this objective.

Solution to this problem may be found in part in the provision of specific housing for the weaned sow. Where sows at the end of lactation are put back into standard dry sow accommodation, there is a tendency for the sow to be forgotten and/or mismanaged. A proportion of them may not be effectively served for an extended period and the average 'empty days' for the herd as a whole rises accordingly. The best management systems include a purpose-built service house to accommodate sows from weaning to 4 weeks post coitum to ensure all sows are pregnant before they move on to the dry sow house. Such a service area should provide close proximity of boars to gilts and sows and also enough space to carry out supervised services. Well designed gates and passages should facilitate swift and easy heat detection of weaned sows.

Temperature of the house is probably not too critical but a constant 18°C should give good results and experience at the National Agricultural Centre herd at Stoneleigh, England has shown that the lighting regimen which works best is a 12 hours on, 12 hours off cycle.

The grouping of sows and gilts within the house may be important. Currently many of the service houses which are being built include individual penning of weaned sows and there are good reasons for this. Bullying problems are removed and individual feeding and observation is straightforward. It may be however that the female-to-female contact in the immediate postweaning period is a necessary part of the sow's behaviour to elicit a good postweaning response and rapid return to fertile heat. The point is undocumented but the authors have observed some commercial situations where group housing of weaned sows (with good general sow management) appears to be associated with a more consistent return to heat, fewer conception problems and apparently reduced weaning to oestrus interval. Hence housing in groups of four or five sows and gilts should give few problems while exploiting these advantages. In addition there is a further advantage of grouping in that the cost is reduced and heat detection is made easier by the stockman observing sows mounting prior to and during oestrus.

Heat detection in fact should begin on day 1 following weaning, probably at this stage with an examination by the stockman, but from day 2 onwards a boar should be run with the groups of females each day or even twice a day to pick up oestrus in any sow the moment it appears. Subsequently service should be closely supervised to ensure coitus is effective and a repeat service carried out 12–24 hours later. Served sows should be thoroughly observed 3 weeks later for any returning heat symptoms and if a negative response is seen this is a reasonable indication of successful service.

Even with the best management there will still in most herds be a proportion of sows and gilts which fail to show a postweaning heat. A decision to cull at some stage must be made but speed is essential. Any sow or gilt remaining unserved for an extended period tends to bring down the average productivity of the whole herd. Varley (1979) has tentatively suggested that sows should be carefully observed up to day 35 post weaning and if a heat is not seen then rapid disposal of the sow should be carried out. After this time the probability of a sow showing heat is low (though not impossible).

Varley and Cole (1976) have also observed in a closely monitored herd that up to 20 per cent of apparently effectively served sows may be found non-pregnant at full term. Presumably there is a large variation between herds in this parameter but if it is general then it represents a large loss of reproductive efficiency. The various pregnancy diagnosis techniques may prove of value in this context and regular checking throughout pregnancy may be a necessary feature of sow management.

CULLING RATES

A survey carried out by the University of Cambridge (Ridgeon, 1974) on a large sample of Eastern Counties farms indicated that the culling rate for sows was of the order of 40 per cent per annum. Unfortunately this figure was not further broken down but the usual reasons for culling apart from ill-health and leg weakness are infertility and sterility. These problems can be listed as: failure to show a postweaning oestrus, silent heat, failure to hold to service, low fertility in terms of litter size. In addition some animals are culled owing to behavioural problems at service.

According to Einarsson (1974), Einarsson and Settergen (1974), Rasbech (1969) and Jones (1967) about a third of breeding stock per annum is replaced and of these a third is culled because of the infertility problems listed above.

Precise information giving specific factors influencing infertility is very limited. Housing systems have been investigated as a possible cause of infertility and Tuinte (1971) has observed a higher total percentage of animals discarded and also a higher percentage culled for anoestrus and silent heat in those housing systems in which straw was used. In addition the culling rate for reproductive malfunction was higher in individually housed animals than in group-housed animals. This last point has some bearing on the present practice of individually tethered sows on slats. At present no comparative data are available to examine the effect of this on reproductive activity.

It has not been established whether high ambient temperature is related to increased evidence of anoestrus and silent heats in sows but the observation has been made in a number of studies in gilts (Jensen *et al.*, 1970; Teague, Roller and Grifo, 1968; Warnick *et al.*, 1965) that high ambient temperature is related to a high incidence of reproductive malfunction.

THE EFFECTS OF INFERTILITY ON SOW OUTPUT

Every sow which returns to heat repeatedly, becomes anoestrous or which is found non-pregnant at expected full term is a drain on the rest of the herd and brings down the mean sow productivity.

There are straightforward mathematical relationships between the level of infertility and annual sow productivity in terms of weaners produced per sow per year. These relationships are illustrated in *Figure 10.3* for three different lactation lengths and it can be clearly seen that with progressive increase in infertility there is a significant decline in sow output. Much of this loss is due to the extremely long

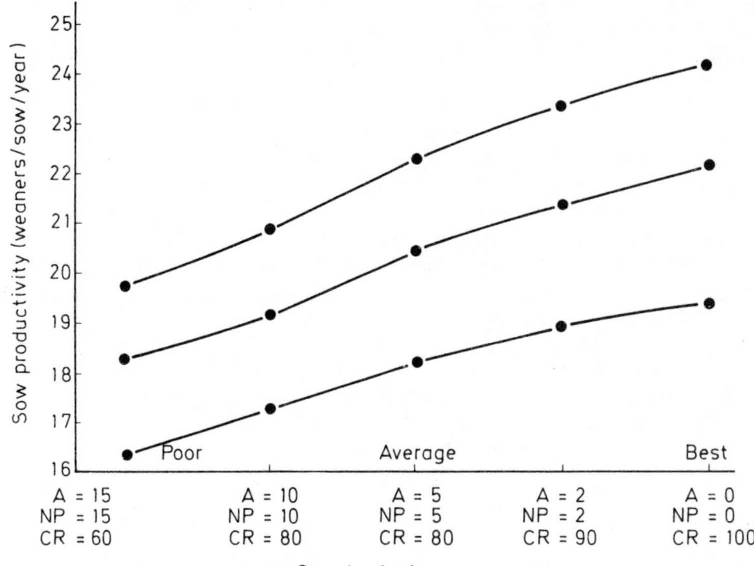

Figure 10.3 The overall effect on sow output of reproductive efficiency. A = percentage of sows anoestrous; NP = percentage of sows found non-pregnant; CR = non-return rate

weaning to conception interval registered for sows and gilts with reproductive problems. This highlights the necessity to manage them correctly to keep these losses to a minimum.

10.4 References

AUMAITRE, A., PEREZ, J.M. and CHAUVEL, J. (1975). *Journées de la recherche porcine en France*, pp. 52–67, Paris, L'Institut Technique du Porc

BROOKS, P.H. and COLE, D.J.A. (1972). *Anim. Prod.* **15**, 259–264

BROOKS, P.H., COLE, D.J.A., ROWLINSON, P., CROXON, V.J. and LUSCOMBE, J.R. (1975). *Anim. Prod.* **20**, 407–412

BURGER, J.F. (1952). *Onderstepoort. J. vet. Res. Suppl.* **2**, 3–218

CLARK, J.R., DAILEY, R.A., FIRST, N.L., CHAPMAN, A.B. and CASIDA, L.E. (1972). *J. Anim. Sci.* **35**, 1216

CRIGHTON, D.B. (1967). In *Reproduction in the Female Mammal* (Ed. by G.E. Lamming and E.C. Amoroso), pp. 222–238. London, Butterworths

DU MESNIL DU BUISSON, F. and SIGNORET, J.P. (1968). *Proc. 6th int. Congr. Anim. Reprod. Paris.* **2**, 1091–1094

DYCK, G.W. (1972). *Can. J. Anim. Sci.* **51**, 135–140

EINARSSON, S. (1974). *25th Annual Meeting E.A.A.P. Commission on Pig Production.* Copenhagen
EINARSSON, S. and SETTERGEN, I. (1974). *Nord. VetMed.* **26**, 576–580
JENSEN, A.H., YEN, J.T., GEHRING, M.M., BAKER, D.H., BECKER, D.E. and HARMON, B.G. (1970). *J. Anim. Sci.* **31**, 745–750
JONES, J.E.T. (1967). *Br. Vet. J.* **123**, 327–339
LEGAULT, C., DAGORN, J. and TASTU, D. (1975). *Journées de la recherche porcine en France*, pp. 42–62. Paris, L'Institut Technique du Porc
LONGNECKER, D.E. and DAY, B.N. (1968). *J. Anim. Sci.* **27**, 924
PALMER, W.M., TEAGUE, H.S. and VENZKE, W.G. (1965). *J. Anim. Sci.* **24**, 541–545
PARLOW, A.F., ANDERSON, L.L. and MELAMPY, R.M. (1964). *Endocrinology* **75**, 365–376
POLGE, C. (1972). *Porc. Br. Soc. Anim. Prod.* **1**, 5–18
RASBECH, N.O. (1969). *Br. Vet. J.* **125**, 599–616
RIDGEON, R.F. (1974). *Pig Management Scheme. Results for 1974.* University of Cambridge, Agricultural Economics Unit
TEAGUE, H.S., ROLLER, W.L. and GRIFO, A.P. Jr. (1968). *J. Anim. Sci.* **19**, 408–411
TUINTE, I.J. (1971). *Mandblad voor de Varkensfokkerij* **33**, 201–204
VARLEY, M.A. (1979). *Pig Fmg* **27(1)**, 52
VARLEY, M.A. and COLE, D.J.A. (1976). *Anim. Prod.* **22**, 21–27
WARNICK, A.C., WALLACE, H.D., PALMER, A.Z., SOSA, E., DUERRE, D.J. and CALDWELL, V.E. (1965). *J. Anim. Sci.* **24**, 89–92

Part two

The male

Introduction

In this section it is intended that the more important aspects of boar management be reviewed in relation to functional anatomy, physiology and behavioural patterns. This is considered to be a necessary part of the book for several reasons:

(1) To complement the detailed account of female reproductive processes.
(2) To provide a review of the present state of knowledge on reproduction in the boar.
(3) To give a background from which to draw conclusions and recommendations on the management and use of boars.

The boar does, of course, contribute 50 per cent of the genetic material for each offspring and, indeed, each individual boar will be expected to contribute 15–25 times as many offspring as the individual sow. Thus it might be said, that in this respect at least, the boar is more important than the sow. In terms of individual contribution this is undoubtedly true and therefore should be borne in mind when considering boar management. It is of little use possessing a herd of sows whose fertility and potential production level is high if the boars to be used are of low libido, inactive or infertile.

Having stated this it is necessary to justify the relative brevity with which this subject is to be dealt. Basically, once maturity has been reached the primary aim of management is to maintain the boar's libido and fertility. This contrasts sharply with the many facets of reproductive efficiency on the female side (preceding chapters). Hence, despite the obvious importance of the boar, the primary objective of breeding herd management must be to maximize output from the sow while maintaining boar fertility. Therefore it is the purpose of this section to provide an account of how boar activity, libido and fertility may be maintained at an optimum level.

Within this framework several areas of importance are apparent. First, the management of the young boar from selection up to, and including, the first few services must be considered in detail. Indeed, the problems of the mature boar may often be traced back to poor management in this period in general and at the first few services in particular. Following maturity the questions to be considered are mainly associated with the frequency of use, nutrition, housing, health status and culling of the boar. Finally, an area of extreme importance which is often neglected, namely the assessment of boar performance, must be considered.

Chapter 11

Puberty in the male

The attainment of puberty in the male is considered to have occurred once free spermatozoa have appeared in the seminiferous tubules and are present in the cauda epididymidis. This represents the time when the male first becomes capable of breeding, although it should be emphasized that fertility is not at a maximum at this stage. Indeed, in a recent experiment Shin *et al.* (1976), working with crossbred boars, found that although puberty was attained at 27 weeks of age the boars' fertility did not reach an acceptable level for breeding until 35 weeks of age. Hence, unlike the situation with the gilt (*see* Chapter 3), we are not primarily concerned with stimulating the attainment of precocious puberty in the boar. Rather, the aim is to allow the boar to mature sexually, and then to increase gradually the number of matings he performs as his level of fertility increases.

Therefore, the information given below is of mainly academic interest since we are not concerned with manipulating the maturation process in the boar. It is important, however, to understand the mechanisms governing maturation, and thus to comprehend the possible influences of management on this development. Furthermore, a knowledge of the development sequence in the male provides an insight into the causes of low fertility in the young mature boar.

11.1 The physiology of puberty

Until recently endocrine mechanisms in the boar have received little attention. Indeed, in 1974 Hafez reported that no reliable figures were available for the plasma levels of any reproductive hormones in the prepubertal boar. Since 1974 the situation has only slightly improved. Several workers have now attempted to measure LH and testosterone in both the prepubertal and mature boar. However, this work is by no means complete, and hence our basic information on endocrine changes in the immature boar is severely limited. For this reason much of the information discussed below relates to species other than the pig. It is recognized that this is not totally satisfactory, but until such time as the information relating to the boar becomes available it is the best estimate that can be given of the endocrine events.

HORMONE LEVELS

The three hormones of importance in male reproduction are LH, FSH and testosterone. LH (also known as ICSH or interstitial cell stimulating hormone in the male) and FSH are, of course, produced by the anterior pituitary gland and are under the control of the hypothalamic releasing hormones. Testosterone is the major gonadal steroid produced by the male, this being synthesized in the Leydig cells of the testis. The testes do, however, produce several other steroids prominent among which are oestrogens and androstenedione. Neither of these steroids is very active, most of the oestrogen appearing in the relatively inactive conjugated form, while the androstenedione (a precursor of testosterone) is only weakly androgenic. Thus, for the purpose of this discussion, the only gonadal steroid which need be considered is testosterone.

FSH

There are no reports of plasma FSH concentration having been measured in the boar during the prepubertal period. However, results obtained in rodents indicate that plasma FSH levels are low in the young male but demonstrate a significant increase as puberty is approached (Ramirez, 1973; Mackinnon, Mattock and Ter Haar, 1976; Selmanoff, Goldman and Ginsburg, 1977).

LH

Reports relating to plasma LH levels prior to puberty in the male tend to be of a contradictory nature. Many results indicate that LH levels are low throughout the prepubertal period, with little variation in plasma concentration (Elsaesser *et al.*, 1976; Mackinnon, Mattock and Ter Haar, 1976; Romanowicz *et al.*, 1977). However, other workers have reported peaks of LH occurring in either the early postnatal period (Ramirez, 1973; Lawson, 1975; Ford and Scanbacher, 1976; Colenbrander *et al.*, 1977) or in the period leading up to puberty attainment (Selmanoff, Goldman and Ginsburg, 1977). Clearly, it is difficult to draw conclusions from such conflicting results. However, it now seems likely that plasma LH concentrations in the male are mostly low throughout development, although very high levels may occur in a small minority of animals (Mackinnon, Puig-Durran and Laynes, 1978).

Testosterone

In the early postnatal male the testis synthesizes and secretes androstenedione in larger amounts than testosterone (Moger, 1975). However, as maturation proceeds a change in testicular steroid metabolism occurs and testosterone output rises relative to androstenedione (Mackinnon, Puig-Durran and Laynes, 1978). This rise

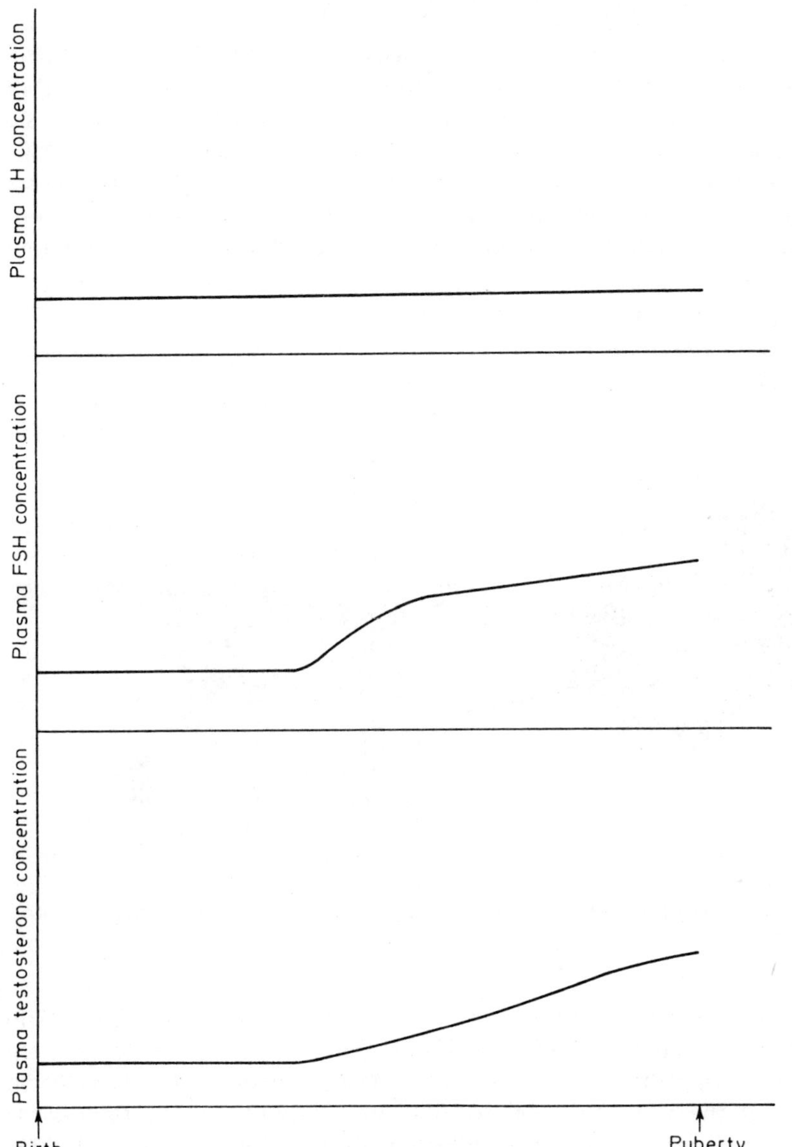

Figure 11.1 Relative changes in plasma LH, FSH and testosterone concentrations during the prepubertal period in the male

in plasma testosterone has been observed in the mouse (Selmanoff, Goldman and Ginsburg, 1977), rat (Mackinnon, Puig-Durran and Laynes, 1978), ram (Lawson, 1975), and boar (Gray *et al.*, 1971; Booth, 1975: Elsaesser *et al.*, 1976; Romanowicz *et al.*, 1977). These hormone changes are shown diagrammatically in *Figure 11.1.*

GROWTH AND DEVELOPMENT OF THE TESTIS

When mature the boar testis consists of long hollow convoluted tubes, known as the seminiferous tubules, surrounded by interstitial tissue. The seminiferous tubules are, in fact, made up of two basic cell types, these being the germ cells (ranging in degree of maturity from primitive germ cells to mature spermatozoa) and the supporting Sertoli cells. The interstitial tissue consists of connective tissue and the androgen-secreting Leydig cells. These structures are shown in *Figure 11.2.*

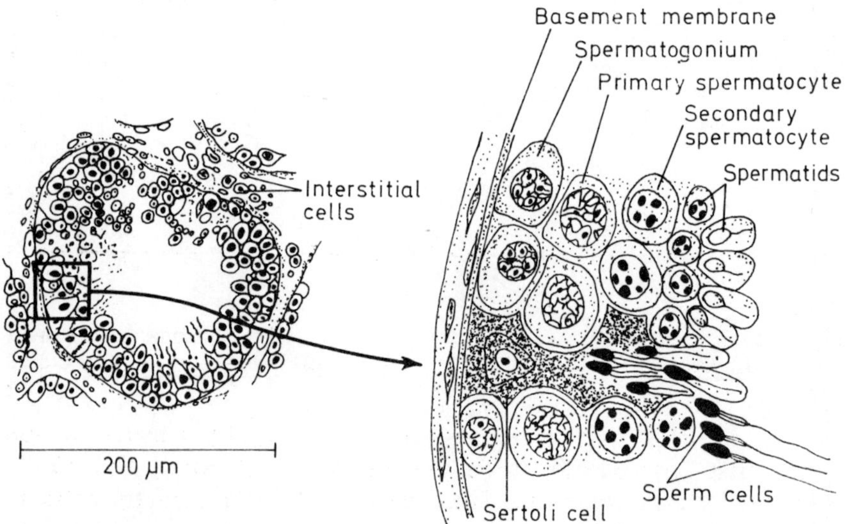

Figure 11.2 Section of testis showing the general morphology (from Bloom and Fawcett, 1968).

At birth, however, the testis is relatively undifferentiated. The precursors of the seminiferous tubules, known as the sex cords, are solid (i.e. they have no central lumen) and consist of primitive germ cells and the precursors of Sertoli cells. In contrast, the interstitial cells (precursors of Leydig cells) are abundant in the testis at this time, but their numbers subsequently diminish during the early post-natal period. It is clear that rapid proliferation of these interstitial cells occurs late in the prenatal period, resulting in interstitial tissue making up a large majority of testicular tissue at birth.

Between birth and the attainment of puberty the relative growth rate of the two tissues (i.e. sex cords and interstitial tissue) changes. Bascom and Osterud (1927) reported that the pig testis at birth consisted of 84 per cent interstitial tissue and only 16 per cent sex cords, whereas by the time maturity had been reached the relative percentages of these tissues had changed to 37.8 per cent and 62.2 per cent respectively (*see Figure 11.3*). Indeed, the data of these

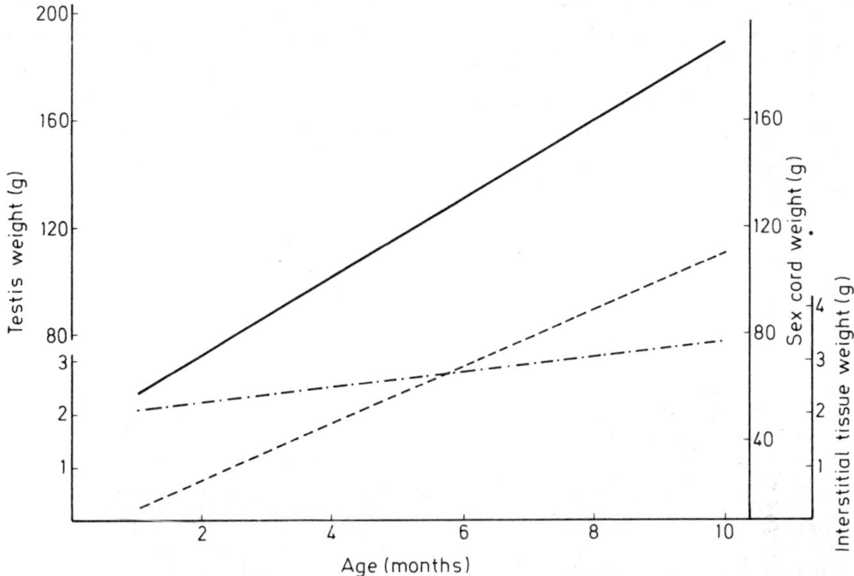

Figure 11.3 Relative changes in testicular tissue weights during development (from Bascom and Osterud, 1927). ——— testis weight; – – – – – sex cord weight; —·—·—· interstitial tissue weight

workers indicate that sex cord weight increases by a factor of 468 in the first year of life, while the weight of interstitial tissue only increases by a factor of 39. Thus, between birth and maturity there is a rapid growth of testicular tissue due to the extension of the sex cords (from a total length of 29 m at birth to 3164 m at 1 year) and, to a lesser extent, the proliferation of the interstitial cells. During this same period the solid sex cords are transformed into the hollow functioning seminiferous tubules of the mature male. Furthermore, the interstitial cells of the prenatal boar develop into the functional androgen-secreting Leydig cells at this time.

Testis growth and development is under the control of the anterior pituitary gland hormones FSH and LH. FSH undoubtedly plays the dominant role and is responsible for the stimulation of seminiferous tubule growth. This may explain the rapid increase in the length and weight of the seminiferous tubules during the prepubertal period, since FSH secretion is increasing at this time (*see Figure 11.1*). In contrast, development of the interstitial tissue does not appear to be dependent on reproductive hormones, although testosterone secretion by the Leydig cells is stimulated by LH.

A recent paper by Esbenshade *et al.* (1979) has described a study of the effects of housing system (indoor vs outdoor) on testes development. The conclusion was made that prepubertal environment from 58 days onwards has no effect on weight gain of testes or testes size. Libido was observed to be higher in boars housed indoors in intensive conditions, however, but this was not associated with a higher output of viable sperm cells at maturity.

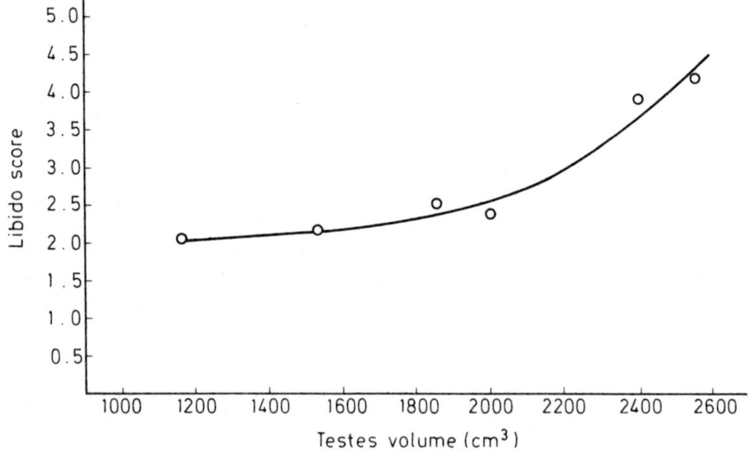

Figure 11.4 The effect of testes volume on libido in boars (from Esbenshade *et al.*, 1979)

Figure 11.4 shows the relationship from Esbenshade *et al.* (1979) between testes volume and libido score as judged subjectively on a scale: 0 = poor, 3 = average, 5 = very aggressive. From this it appears that increased testes size is associated with enhanced libido.

11.2 Spermatogenesis

The term spermatogenesis is used here to describe the developmental process whereby a primitive germ cell is converted into a mature spermatozoon. This process will be dealt with in detail in Chapter 12, but must also be mentioned here as it has an important bearing on the developmental events leading up to puberty attainment. We have already defined puberty as being the time when mature spermatozoa first appear. Hence, puberty must be the time when the first wave of spermatogenesis is complete.

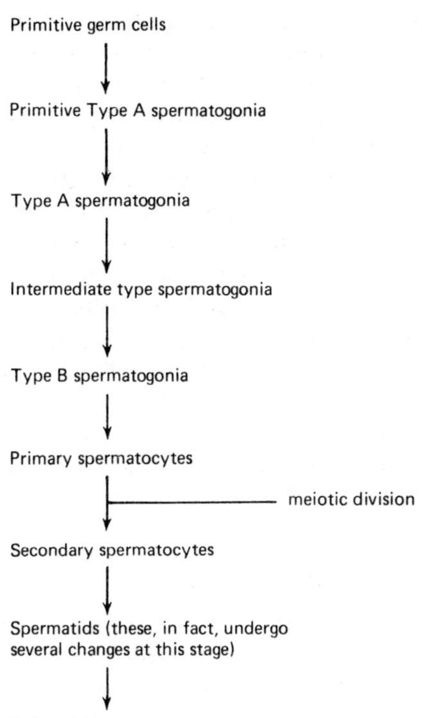

Figure 11.5 Developmental stages in the transformation of primitive germ cells to mature spermatozoa

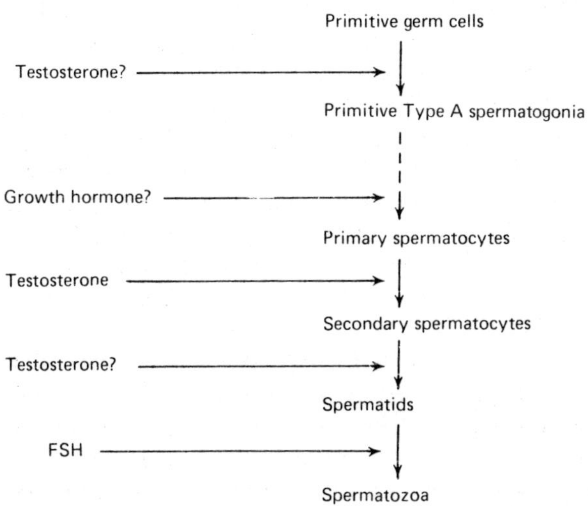

Figure 11.6 Hormonal involvement in spermatogenesis. ? indicates that this hormone may be involved with this particular step. However, the step may not require the presence of hormones

The steps involved in spermatogenesis are shown in simplified form in *Figure 11.5*. The time taken for this sequence of events to occur is fixed for each species. In the case of the boar, this time period is approximately 34 days (Swierstra, 1967). Thus, it would seem logical to conclude that the boar would reach puberty about 34 days after the first primitive germ cells were initiated into the spermatogenetic sequence. However, this is not the case. Kim and Lee (1975) have reported that spermatogonia first appear in the testis of the boar at 2 months, spermatocytes at 3 months and spermatozoa at 5 months. These results indicate that there is a lack of stimulation during the final stages of spermatogenesis in the prepubertal boar. This delay is probably attributable to the low levels of testosterone and FSH apparent in the male until the period immediately preceding puberty. It is now recognized that these two hormones are necessary for the maturation of spermatozoa from the primary spermatocyte stage through to the mature spermatozoa (*Figure 11.6*).

11.3 Mechanisms controlling puberty attainment

It is clear from the foregoing discussion that the attainment of puberty in the male is primarily dependent on an increased secretion of testosterone and FSH, facilitating the final steps of spermatogenesis. Therefore, the purpose of this section is to consider the mechanisms whereby the circulating levels of these hormones are raised during the prepubertal period.

In the early postnatal period the circulating levels of FSH, LH and testosterone are low. This is due to the testosterone having a negative feedback effect on the secretion of the gonadotrophins. However, at some stage in the period prior to puberty the hypothalamus becomes less sensitive to this negative feedback effect. This change in the 'gonadostat' has previously been described in the female (Chapter 3). In the case of the male the 'gonadostat' change appears to have a differential effect, in that the change in sensitivity appears to be effective for FSH but not LH (Mackinnon, Puig-Durran and Laynes, 1978). Thus, once this change in sensitivity occurs there is a rise in plasma FSH levels, since the inhibitory influence of testosterone is effectively reduced (*see Figure 11.1*).

However, it has been stated above that both FSH and testosterone are necessary for spermatogenesis to be completed. Therefore, an increase in testosterone levels must also occur during this period. In fact, testosterone secretion is directly influenced by LH, and therefore we must consider what is happening to this hormone during the prepubertal period. *Figure 11.1* shows that plasma LH concentration remains low throughout the period prior to puberty, and therefore it must be concluded that a change of sensitivity to LH must occur at the gonadal level. Recent experiments have shown that increased circulating levels of FSH promote the development of LH receptors in the testis (Odell and Swerdloff, 1976; Mackinnon, Puig-Durran and Laynes, 1978). Thus, the increased plasma FSH levels result in the testis becoming more sensitive to LH, and hence, an increased rate of secretion of testosterone.

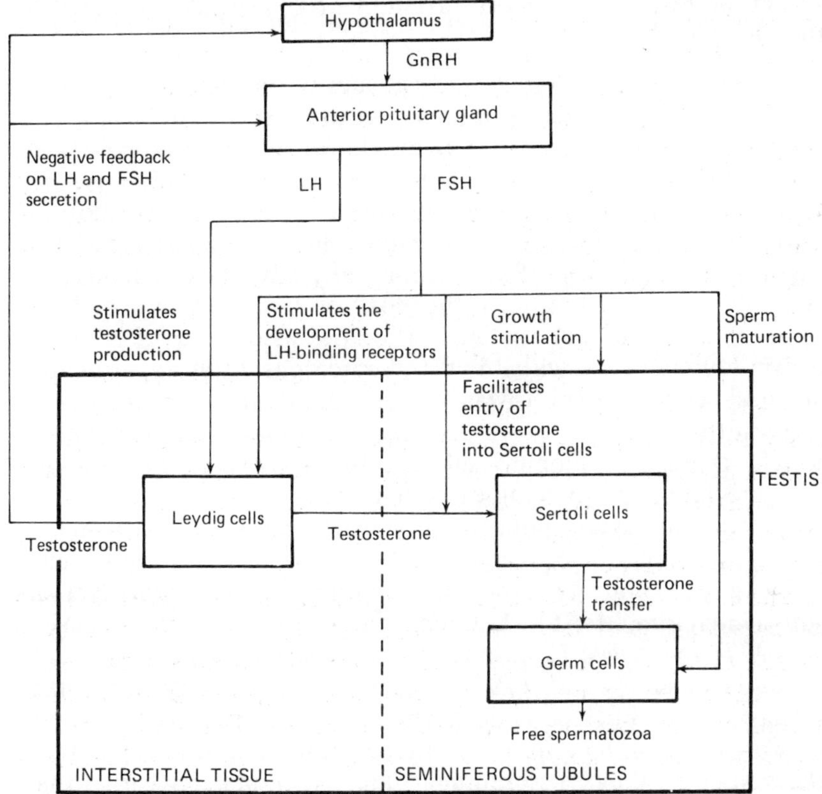

Figure 11.7 The interaction of FSH, LH and testosterone in the prepubertal male

This control mechanism explains how plasma levels of both FSH and testosterone become elevated during the prepubertal period. Furthermore, since these two hormones promote the final stages of maturation of the spermatozoa, this mechanism may be used to explain the attainment of puberty in the male. However, one further factor must be considered before the picture is complete. This is the mechanism whereby the testosterone reaches its site of action, namely the germ cells. It now seems likely that the testosterone initially binds with medium-affinity binding proteins in the Sertoli cells. This binding is facilitated by the presence of FSH. Once in the Sertoli cells the testosterone is rapidly transferred to the germ cells, since these contain high-affinity binding proteins. Hence, the testosterone reaches its site of action via the Sertoli cells and with the aid of FSH.

The interactions of FSH, LH and testosterone are summarized in *Figure 11.7.*

11.4 Factors influencing puberty attainment and subsequent fertility in the young boar

It has previously been stated that the aim when rearing boars is to allow them to attain puberty rather than to stimulate the precocious attainment of sexual maturity. The reason for this is that the fertility of the young boar is relatively low immediately after puberty, and does not reach a maximum until at least the age of 12–18 months. Thus, there is little advantage to be gained, in terms of mating ability, by reducing pubertal age. Furthermore, any advantage that may be gained must be offset against the cost of the treatment and the possible enhancement of mating difficulties associated with young boars of low body weights. For these reasons little experimental effort has been directed towards methods of inducing precocious puberty in the boar. The few results that are available are given below, although it must be emphasized that the sparsity of data on this subject precludes the drawing of conclusions.

Boars normally attain puberty between 5 and 8 months (Kim and Lee, 1975; Leman and Rodeffer, 1976), when they will weigh 70–120 kg. This appears to be more closely related to boar age than weight (Einarsson, 1975), although one recent experiment (Kim *et al.*, 1977) does suggest that nutrition may influence puberty attainment. Indeed, Kim *et al.* (1977) observed that puberty attainment was delayed by as much as 47 days when the food intake of the young boar was reduced (*Table 11.1*). In addition, the results of these workers indicate that crossbred boars may reach puberty earlier than their purebred counterparts, an observation supported by the results of Sellier, Dufour and Rousseau (1973). Thus, it seems likely that puberty may be delayed by reducing food intake and stimulated in a crossbred boar.

Table 11.1. The effects of genotype and feed level on puberty attainment in the boar

		Age at puberty (days)
Purebred boars –	standard diet	215
–	70% standard diet	262
Crossbred boars –	standard diet	179
–	70% standard diet	209

From Kim *et al.* (1977).

However, the above results do not truly relate to the induction of precocious puberty. With one exception (Oettingen, Smidt and Merkt, 1973), no attempt has been made to induce the early attainment of sexual maturity in the boar. The experiment of Oettingen, Smidt and Merkt (1973) was carried out to elucidate the influence of exogenous hormones on sexual development in the boar. Boars of 55–60 days of age were treated with either FSH, LH, FSH + LH, PMS, HCG or PMS + HCG for a 30-day period. At the end of this period it was observed that all treatments had stimulated an increase in testis growth. In addition, the spermatogenic sequence was clearly further advanced in those boars treated with PMS, HCG or PMS + HCG.

Fully developed spermatozoa were present in 50 per cent of PMS + HCG treated boars. However, although these results confirm the roles of FSH and testosterone (induced by LH) in the development of spermatozoa, it is unlikely that such treatments will have any practical application. The cost of treatment over a 30-day period together with the obvious problems of attempting to use a very young boar for mating, would undoubtedly prohibit the application of such a system in practice.

Once puberty has been reached the boar will be capable of producing mature spermatozoa and, therefore, capable of achieving successful copulation and fertilization. However, as has been previously mentioned, the fertility of the boar is not at a maximum immediately following puberty attainment. Fertility is relatively low at this time, building up to a maximum over a period of 1–2 years. This statement is confirmed by the results obtained in a recent study carried out by Cerovsky (1977). This worker, studying boars from 7 to 36 months of age, observed that the largest ejaculate volume occurred at 29 months, while the smallest volume was collected at 7 months. Furthermore, both sperm concentration and the number of sperm per ejaculate were lowest at 7 months, the highest sperm per ejaculate occurring at 32 months. This appears to be due to differences in the number of spermatozoa undergoing the final stages of maturation in the epididymis in boars of different ages (Egbunike *et al.*, 1976; Holzler *et al.*, 1977). Thus, although the young boar is fertile in the period immediately following puberty attainment, its actual level of fertility is low compared with that of more mature boars (i.e. boars of 2–3 years).

A final area which is worthy of consideration in relation to the fertility of the young boar is the influence of age on pheromone production. This is of importance since the pheromone certainly plays a role in the induction of the 'standing heat reflex' (*see* Chapter 13) and, possibly, in the stimulation of precocious puberty in the gilt (*see* Chapter 3). The pheromone, a combination of 16-androstenes (Patterson, 1968b; Reed, Melrose and Patterson, 1974; Booth, 1975), is found in the testes, fat, submaxillary glands, saliva and sweat glands of the boar (Patterson, 1968a,b; Stinson and Patterson, 1971; Booth, 1975). Although it was generally thought that this pheromone was a metabolite of testosterone, the results of Ahmad and Gower (1968) indicate that it is more likely to be formed from progesterone, possibly under the influence of testosterone. The rate of pheromone production is low in the prepubertal boar, but increases rapidly as maturity is approached. Indeed, the results of Claus, Hoffman and Karg (1971) suggest that the circulating levels of pheromones closely parallel those of testosterone (*Table 11.2*). Hence, it is likely that

Table 11.2. The influence of age and weight on plasma levels of testosterone and pheromones

Sex	*Liveweight* (kg)	*Age* (days)	*Plasma pheromone** *concentration* (ng/ml)	*Plasma testosterone concentration* (ng/ml)
Intact male	100	175	6.0	9.6
	120	217	17.6	15.0
	140	229	22.3	16.1
Intact female	100	177	0.8	0.4
	120	229	2.0	1.4
	140	238	2.0	1.9
Castrate male	100	175	2.7	1.5
	120	217	1.7	2.1
	140	236	1.3	2.4

* The actual pheromone measured in this experiment was 5α-androst-16-en-3-one.
From Claus, Hoffman and Karg (1971).

by the time the young boar attains puberty the rate of pheromone production will be adequate to induce the 'standing heat reflex' in oestrous sows. However, it should be pointed out that little is known about the rate of pheromone production in mature boars. Therefore, the ability of the young mature boar, relative to that of the older boar, to induce the 'standing heat reflex' in oestrous females cannot be accurately assessed.

11.5 Conclusions

The foregoing discussion has emphasized that the fertility of the young boar is of greater importance than the actual timing of puberty attainment. It has also been stated that the aim when rearing boars is to 'allow' them to attain sexual maturity. However, it is also clear that no benefit is to be gained by delaying the onset of sexual maturity. Therefore, the possible adverse effects of undernutrition, or more precisely low plane feeding, should be avoided. On the other hand, overfeeding can lead to obese, clumsy boars whose performance at copulation will be inferior to that of fitter boars. Hence, the aim when rearing the boar, and indeed throughout his active life, should be to keep him 'fit not fat'.

At the time of puberty attainment the fertility of the boar is undoubtedly low. However, this does increase rapidly in the subsequent months. Although maximum fertility appears to occur at about 2½ years, it is clear that the boar's fertility at 1 year is sufficiently high to allow regular matings (i.e. 2 or 3 double services per week). Thus, it would seem advisable to increase gradually the number of matings that a young boar performs, starting at about 6–8 months and building up to the workload of a mature boar by 1 year.

Finally, it should be added that several other aspects of boar management, not dealt with in this chapter, are discussed elsewhere. In particular, factors influencing boar fertility are reviewed in Chapter 12, and the influence of rearing and housing conditions on mating ability is presented in Chapter 13.

11.6 References

AHMAD, N. and GOWER, D.B. (1968). *Biochem. J.* **108**, 233

BASCOM, K.F. and OSTERUD, M.L. (1927). *Anat. Rec.* **37**, 63

BLOOM, W. and FAWCETT, D.W. (1968). In *Textbook of Histology*, 9th edn. Philadelphia, W.B. Saunders and Company

BOOTH, W.D. (1975). *J. Reprod. Fert.* **42**, 459

CEROVSKY, J. (1977). *Anim. Breed. Abstr.* **45**, 2867

CLAUS, R., HOFFMAN, B. and KARG, H. (1971). *J. Anim. Sci.* **33**, 1293

COLENBRANDER, B., DRUIP, T.A.M., DIELEMAN, S.J. and WENSING, C.J.G. (1977). *Biol. Reprod.* **17**, 506

EGBUNIKE, G.N., HOLTZ, W., ENDELL, W. and SMIDT, D. (1976). *Anim. Breed. Abstr.* **44**, 3827

EINARSSON, S. (1975). Cited in Leman and Rodeffer (1976)

ELSAESSER, F., ELLENDORFF, F., POMERANTZ, D.K., PARVIZI, N. and SMIDT, D. (1976). *J. Endocr.* **68**, 347

ESBENSHADE, K.L., SINGLETON, W.L., CLEGG, E.D. and JONES, H.W. (1979). *J. Anim. Sci.* **48**, 246

FORD, J.J. and SCANBACHER, B.D. (1976). *58th Annual meeting of the Endocrine Society.* San Francisco, California

GRAY, R.C., DAY, B.N. LASLEY, J.F. and TRIBBLE, L.F. (1971). *J. Anim. Sci.* **33**, 124

HAFEZ, E.S.E. (1974). *Reproduction in Farm Animals,* 3rd edn. Philadelphia, Lea and Febiger

HOLZLER, J., WETTEMAN, R.P., JOHNSON, R.K. and WELTY, S. (1977). *Anim. Breed. Abstr.* **45**, 2869

KIM, J.K. and LEE, Y.B. (1975). *Korean J. Anim. Sci.* **17**, 294

KIM, J.K., SUH, G.S., SUL, D.S., KIM, V.B. and LEE, Y.B. (1977). *Anim. Breed. Abstr.* **45**, 6655

LAWSON, R.A.S. (1975). *J. Reprod. Fert.* **43**, 379

LEMAN, A.D. and RODEFFER, H.E. (1976). *Vet. Rec.* **98**, 457

MACKINNON, P.C.B., MATTOCK, J.M. and TER HAAR, M.B. (1976). *J. Endocr.* **70**, 361

MACKINNON, P.C.B., PUIG-DURAN, E. and LAYNES, R. (1978). *J. Reprod. Fert.* **62**, 401

MOGER, W.H. (1975). *J. Endocr.* **67**, 135

ODELL, W.D. and SWERDLOFF, R.S. (1976). *Recent Progr. Hormone Res.* **32**, 245

OETTINGEN, U. von., SMIDT, D. and MERKT, H. (1973). *Anim. Breed. Abstr.* **41**, 4491

PATTERSON, R.L.S. (1968a). *J. Sci. Fd Agric.* **19**, 31

PATTERSON, R.L.S. (1968b). *J. Sci. Fd Agric.* **19**, 434

RAMIREZ, V.D. (1973). In *Handbook of Physiology* (Ed. by S.R. Geiger). *Section 7, Volume II, Part 1.* Washington, D.C., American Physiological Society

REED, H.C.B., MELROSE, D.R. and PATTERSON, R.L.S. (1974). *Br. Vet. J.* **130**, 61

ROMANOWICZ, K., STUPNICKI, R., BARCIKOWSKI, B., MADEJ, A. and WALACH, M. (1977). *Anim. Breed. Abstr.* **45**, 1995

SELLIER, P., DUFOUR, L. and ROUSSEAU, G. (1973). *Anim. Breed. Abstr.* **41**, 1191

SELMANOFF, M.K., GOLDMAN, B.D. and GINSBURG, B.E. (1977). *Endocrinology* **100**, 122

SHIN, W.J., LEE, Y.B., KIM, J.K. and KIM, C.K. (1976). *Anim. Breed. Abstr.* **44**, 251

STEINBERGER, E. (1971). *Physiol. Rev.* **51**, 1

STINSON, C.G. and PATTERSON, R.L.S. (1972). *Br. vet. J.* **128**, 41

SWIERSTRA, E.E. (1967). *J. Anim. Sci.* **26**, 953

Chapter 12

Fertility in the male

Considerable research effort in recent years has been directed at the control and understanding of infertility in the female and as yet our knowledge in this area is far from complete. Our appreciation of factors affecting fertility level in the boar, however, is at an even earlier stage of development. The old adage that 'the boar is half the herd' still rings true and in a practical situation, a herd suddenly struck with a spate of low fertility on the male side can have catastrophic effects on the smooth running of the pig unit for many months to come. One of the main problems is that unlike the sow, where infertility is quickly seen as either a mating problem or as reduced litter size, in the boar it is possible to continue using an infertile stock boar for a long time before it becomes obvious that all the returns to service are due to this one animal. By this time the damage is done and there may be a large number of sows yet to return, or perhaps worse, a large number of sows carrying very small litters. Obviously for a large herd the need for a very efficient boar recording and monitoring system is paramount.

Libido is linked to fertility, in that many of the causes of reduced fertility also induce poor libido. This relationship has some bearing on the fact that boars with a high libido and high fertility tend to be overworked and as a consequence show reduced reproductive ability.

It is not proposed to review here the whole spectrum of factors affecting fertility in the male, but it is hoped to cover some of the more important practical issues and areas which should yield significant progress in the near future as a result of current and future investigational work. For the reader looking for a more detailed account, particularly of the role of pathogenic organisms in infertility a very comprehensive review is given by Wrathall (1975).

12.1 The assessment of fertility

It is often difficult in practice to state categorically whether a particular situation is 'infertility' or not, and there is a very grey area between maximum sperm production and viability on the one hand and total termination of sperm output on the other. Within this middle band a boar might continue for some time producing semen at something less than peak efficiency before being observed as having an infertility problem. How then do we measure level of fertility? There are a number of ways but most of them focus on the total sperm output over a period or at one ejaculate or the sperm concentration per given volume of semen. At the farm level fertility of individual boars is assessed by the prolificacy of sows and by assessing the percentage of returns to service for sows put to the various boars. There are many difficulties when it comes to making culling decisions. It has to be decided what level of prolificacy (or returns) is considered to constitute infertility. There are also many other factors which have nothing whatsoever to do with the boar which could be responsible for temporary loss in reproductive efficiency in a group of sows or gilts. Hasty decisions are therefore often erroneous decisions and especially where a boar is of high genetic merit (and hence high cash value), a veterinary check should be made to give a more precise measure of fertility including an assessment of sperm quantity and quality.

12.2 Components of semen quality

Table 12.1 gives the chemical characteristics of an average sample of boar semen. It must be emphasized, however, that there is large variation around these typical values. Where these values reach the

Table 12.1. The chemical characteristics of boar semen

Property	*Mean* (mg/100 ml)	*Range* (mg/100 ml)
pH	7.5	7.3 – 7.8
Water	95	94 – 98
Sodium	650	290 – 850
Potassium	240	80 – 380
Calcium	5	2 – 6
Magnesium	11	5 – 14
Chloride	330	260 – 430
Fructose	13	3 – 50
Sorbitol	12	6 – 18
Citric acid	130	30 – 330
Inositol	530	380 – 630
Glycerophosphorylcholine		110 – 240
Ergothioneine		6 – 23
Protein	3700	

From Hafez (1976).

Table 12.2. Boar semen and its physical characteristics

Property	*Range*
Volume (ml)	150 – 300
Sperm concentration (10^6/ml)	200 – 300
Total sperm/ejaculation (10^9)	30 – 60
Total sperm/week (10^9)	100 – 150
Motile sperm (%)	50 – 90
Morphologically normal sperm (%)	70 – 90

From Hafez (1976).

extremes in a particular sample of semen, it may be that the fertilizing ability of this sample is impaired to a varying extent. *Table 12.2* gives the physical characteristics of boar semen.

The values given in *Table 12.2* are again 'normal range' values. Infertility is usually associated with values below those given here.

One of the major factors in fertility appears to be the motility of individual sperm cells and it is standard practice at A.I. centres simply to view a sample of ejaculate under the microscope to assess subjectively the wave motion and to rate the sample accordingly. This provides a somewhat crude but effective measure of the fertility level of the particular sample. Techniques to make this type of assessment more objective have been investigated and probably the best method is the use of the impedance bridge which measures the rate of change of electrical resistance of a sperm suspension. The impedance change frequency (I.C.F.) is correlated with the activity of dense sperm suspensions. This technique is now used by the larger bull A.I. centres but the cost of the equipment is probably prohibitive to its use by smaller pig A.I. stations. In all probability a simple microscopic subjective assessment is the most cost-effective way of assessing semen for motility. Chemical analysis has also been examined as a possible indicator of fertility and in particular the rates of fructolysis and respiration are correlated with the motility of sperm. The correlation coefficients involved however are not high enough to facilitate an accurate evaluation of individual semen samples.

Sperm concentration is another way of assessing overall semen quality. This can be measured in a number of ways. The most accurate involves the use of a haemocytometer to count the numbers of sperms actually in a given sample. This technique is time-consuming, however, and so the use of a photometer or colorimeter gives a reasonable result far more quickly. This works on the principle that the number of sperm per millilitre affects the optical density of the sample. It is thus quite easy to calibrate a standard laboratory photometer against known samples to assess unknown samples for sperm concentration (Foote, 1972).

Nearly all samples of semen contain some abnormally formed sperm cells. Abnormalities such as bent tails, acrosomal loosening, detached heads, incorrect attachment of tail, cytoplasmic droplet still attached and others are all observed in a normal sample of semen. It is only when the percentage of abnormal sperm cells rises above 20 per cent that this will affect fertility. In practice this percentage can be estimated using simple microscopic observations together with the correct staining techniques. It is interesting to note that Saacke (1972) has concluded that acrosomal changes are more highly correlated with fertility than is sperm motility. As yet we do not have the techniques to assess accurately and cheaply acrosomal normality in a sample of semen.

12.3 Factors affecting male fertility

GENOTYPE

Generally speaking, the heritability of characteristics associated with male fertility is medium to low, which suggests a large environmental component of the phenotypic variation observed. Precise figures for the boar are not available at the present time but Abadia (1972) has estimated that the heritability for sperm concentration in dairy bulls is 0.28 and motility has a heritability of 0.23. Inbreeding also adversely affects the expression of male fertility and heterosis enhances the effect (Abadia, 1972). The biggest effect of heterosis on fertility appears to be the reduction in the percentage of abnormal sperm cells. It would be possible to make significant progress in a selection programme in male fertility characteristics but the annual selection response would be slow and indeed in our national pig improvement programmes there are far more important selection objectives which can yield a faster economic response. It has also been shown that the boar carries genes which influence fertility of daughters (Ilancic and Pandza, 1973; Strang, 1970; Simovic and Milojic, 1968; Milojic and Rasajski, 1975) but it is not established whether the correlation between boar fertility and daughter fertility is high.

There are significant between-breed differences in sperm production and generally the larger breeds such as the Yorkshire and Large White tend to produce a greater volume of semen per ejaculate and greater numbers of sperm cells over a period, although it is not clear how mature size affects sperm concentration.

It has been observed (Johnson and Gerrits, 1977) that highly selected lines where the rate of gain, feed conversion efficiency and backfat thickness have all been improved rapidly over a very short period are associated with poor boar fertility compared to unselected lines.

It is probable, therefore, that boars resulting from highly selected lines become imbalanced physiologically and as a consequence reproductive performance is reduced.

THE EFFECT OF AGE

After puberty occurs at between 5 and 8 months the number of sperm and the volume of ejaculate increase until the boar reaches 18 months (Leman and Rodeffer, 1976). The ejaculate at this time may consist of between 20 and 80 $\times$ 10^9 sperm in 200–400 ml of semen. This level of sperm production is maintained until a gradual decline sets in after year 5 of the boar's life. There is probably a tremendous variation in practice in the time individual boars reach reproductive senescence. Some boars may continue to serve sows effectively for a very prolonged time while others end their working lives at around 7–8 years. In commercial practice, of course, there is little need to have boars in use beyond 3–5 years. Indeed since the advent of hybrid breeding programmes and rapid generation turnover, it has become imperative for a producer to replace his working stock boars frequently to keep up with genetic progress.

Little is known about the effect of age on the percentage of abnormal sperm cells and it is likely that although a boar can maintain its libido and semen output for many years, the percentage of abnormal sperm cells may rise considerably to reduce the fertility of the boar.

THE EFFECTS OF NUTRITION

The boar normally has the ability to carry on producing quantities of viable sperm cells within a wide range of nutritional fluctuations. It has not been properly established what the limits of this range are but for most breeds a daily energy intake of between 25 and 35 MJ digestible energy should maintain body condition and hence fertility in most boars. We do not have specific rations for boars: working stock boars are invariably fed a standard 'low protein' dry sow feed with acceptable results. Deficiencies of specific vitamins and minerals may be responsible for loss of fertility but in commercial practice this situation (hopefully) never arises. More often than not feeding scales of boars are left entirely to the discretion of the stockman and the aim should be to modify these when necessary to keep the boars in adequate body condition without allowing them to put on excess body fat. Boars carrying too much fat tend to show reduced libido.

ENVIRONMENTAL EFFECTS

Swiestra (1970) has observed that high ambient temperature is associated with reduced sperm motility and concentration thus reducing fertility. This observation is substantiated by the work of Christenson (1973) who has demonstrated that there may in fact be a time lag of 15 days from the exposure of boars to high ambient temperature to the actual drop in fertility level. Normal fertility is then restored some 60 days after the initial exposure as a result of the spermatic cycle being 34 days and sperm cells residing in the epididymis for about 10 days. It thus appears that any short term exposure to high temperature may have quite long term effects on fertility.

Severe spermatogenic damage occurs when testicular temperature reaches 40.5 °C (Mazzari, 1969), and as body temperature is closely related to testicular temperature any disease associated with high body temperature may damage testicular tissue directly and therefore induce infertility.

Cold temperature conditions on the other hand do not generally adversely affect semen quality or fertility (Swiestra, 1970).

These findings are corroborated by the practical experience of pig producers during the United Kingdom summer drought of 1976 when extremely high temperatures were recorded. Most ventilation systems could not cope with the situation and as a result much infertility occurred.

Seasonal variations in sperm production probably do exist but this is a very poorly documented area. Tentatively it might be suggested that changing light patterns affect the production of FSH and hence testosterone and so cause variations in sperm concentration. If this effect is evident it is likely that it is a small effect and of no real practical significance.

HEALTH STATUS

The effects of health status and male fertility are well documented and it is not proposed to review these effects here (*see* Wrathall, 1975; Leman and Rodeffer, 1976).

Physical abnormalities of the development of the penis can lead to infertility and in many cases can be corrected by surgery or by sexual rest. There is also a wide range of pathogenic organisms which are commonly localized in the genital organs of boars and which from time to time flare up causing temporary or permanent infertility problems. It is probably desirable therefore to have a regular veterinary inspection of boars in use, particularly for those of high genetic merit.

FREQUENCY OF USE

Undoubtedly one of the biggest single factors causing temporary loss of fertility in some boars is over-use, or too frequent collection in the case of A.I. boars. Where a stockman has a choice of boars, there is always the tendency for him to use the boar with the strongest libido and easiest handling characteristics, although these are not always linked. In this situation these highly satiated males may apparently be working to the stockman's satisfaction but the net result is poor conception rates.

The boar-to-sow ratio is obviously linked to frequency of use. Rashbech (1969) has stated that young boars should not be used more than once daily. The authors of the present work, however, feel this is probably too high and would suggest that a boar should serve two sows per week. Assuming double service is used, this is four services per 7 days. Hence for every 100 sows kept, assuming a farrowing index of 2.3, five sows will be served every week and therefore three boars would be needed. To take account of temporary infertility problems which arise, an extra boar would be needed as an insurance giving a ratio of 1 boar to 25 sows. It is possible in the short term to operate with fewer boars but in the long term this will almost certainly lead to loss in production due to falling conception rates. Extra boar capacity is also desirable to allow for the 'acclimatization' period for young boars being brought into use.

12.4 Conclusions

According to Leman and Rodeffer (1976), boars with unsatisfactory conception rates are frequently young, heavily muscled boars that are mated excessively while under 8 months of age. This statement illustrates precisely the practice to be avoided if fertility in boars is to be maintained at a high level. In other words, it is essential to introduce young boars carefully and gradually to their full working role and this should culminate in strict adherence to the rules concerning frequency of use laid out above. There is evidence that early animal management in the peripubertal phase may significantly influence the lifetime performance of boars, adversely affecting as well as fertility the other important aspect, libido. Aberrant sexual behaviour can result from isolation during early development and from absence of early heterosexual experience (Dzuik, 1971). It is apparent also that although this fact has been appreciated for some time, the ideal ethological pattern for young boars to experience in early life to elicit the maximum expression of both fertility and libido over the whole lifespan is still not established.

Service houses designed specifically for housing boars and weaned sows should help to reduce the problems due to infertility in boars by facilitating a closer monitoring system for individual males. Simple but efficient recording schemes for quickly detecting boars with a significantly higher proportion of returns than average should be put into effect and where a problem boar is identified in this way, he should rapidly be withdrawn from service until a veterinary check has given him the all-clear. In many cases steroid hormone therapy may rectify the temporary fault but even with the best of management systems, the elimination of non-productive boars is a constant process. Field investigations of unsatisfactory boars has shown that culling of boars is due to the following reasons:

(1) unsatisfactory conception rates
(2) aberrant sexual behaviour
(3) insufficient libido
(4) penis problems
(5) locomotor dysfunction
(6) excessive aggression
(7) old age
(8) ill health

It has not been demonstrated what the relative importance of each of these factors is in the overall culling of boars, but one might expect that because unsatisfactory conception rate is relatively difficult to detect, particularly in the larger herd, then this percentage is not as high as it ought to be.

It may be that there is a tremendous number of boars which exist as 'passengers', and in practice the sow or gilt gets blamed for far more infertility than she is responsible for.

12.5 References

ABADIA, D. (1972). Ph.D. Thesis. Colorado State University, Fort Collins, Colorado

CHRISTENSON, R.K. (1973). *Proceedings of the George A. Young Conference on Advances in Swine Reproduction, Lincoln, Nebraska.* p.3

DZUIK, P.J. (1971). *Proc. Am. Pork Congr.*, p. 159 Des Moines, Iowa

FOOTE, R.H. (1972). *N.A.A.B. Proc. 4th Tech. Conf. Anim. Reprod. A.I.*, pp. 57–61

HAFEZ, E.S.E. (1976). *Reproduction in Farm Animals.* Philadelphia, Lea and Febiger

ILANCIC, D. and PANDZA, F. (1973). *Stcarstuo* **27(1–2)**, 43–49

JOHNSON, L. and GERRITTS, R.J. (1977). *J. Anim. Sci.* **26**, 946

LEMAN, A.D. and RODEFFER, H.E. (1976). *Vet. Rec.* **98**, 457–459

MAZZARI, G. (1969). PhD Thesis, University of Paris

MILOJIC, M. and RASAJSKI, M. (1975). Uticaj nerastoua na plodnost. Nauka i praska. Belgrade

RASBECH, N.O. (1969). *Br. vet. J.* **125**, 599

SAACKE, R.G. (1972). *N.A.A.B. Proc. 4th Tech. Conf. Anim. Reprod. A.I.*, pp. 22–28

SIMOVIC, B. and MILOJIC, M. (1968). *Proc. 6th int. Congr. Anim. Reprod. A.I.*, pp. 1341–1343

STRANG, G.S. (1970). *Anim. Prod.* **12**, 225–233

SWIESTRA, E.E. (1970). In *Effect of disease and stress on reproductive efficiency in swine* (Ed. by M. Lucas and J.F. Wagner), p. 8. University of Nebraska Coop. Ext. Service

WRATHALL, A.E. (1975). *Reproductive Disorders of Pigs.* Slough, England, Commonwealth Agriculture Bureau

Chapter 13

Mating behaviour

Behaviour at the time of mating is an area of study which is recognized to have important practical implications, and has therefore received considerable attention in recent years. There is an obvious necessity to develop normal patterns of behaviour in the young animal, both male and female, in order to facilitate maximum performance at mating.

It is difficult to estimate the influence of behavioural patterns at, or around, the time of oestrus on reproductive efficiency. However, it is clear that many oestrous females do not achieve copulation when taken to the boar for mating. Indeed these females may contribute significantly towards a herd's 'empty days' and therefore provide an adverse influence on overall reproductive efficiency.

The problems associated with mating failure from behavioural abnormalities appear to be most acute in the gilt. This may be largely due to lack of previous mating experience, although the problem is certainly exacerbated by the absence of boar contact prior to mating.

Table 13.1. The influence of pre-mating boar contact on mating rate* in the gilt

	Mating rate (%)
Gilts with previous boar experience	90.0
Gilts without previous boar experience	63.9

* Mating rate is a term employed to describe the percentage of oestrous gilts that actually achieve copulation with a boar.
From Hughes and Cole (1978).

Thus, in a recent experiment Hughes and Cole (1978) observed that mating rate, which is the percentage of females served to those in oestrus was significantly depressed if gilts were reared in isolation from a boar. The results of this experiment (*Table 13.1*) demonstrate that exposure to a boar over the period prior to mating may considerably enhance the gilt's ability to achieve successful copulation.

This is not, however, the only example of management procedure influencing behavioural patterns which, in turn, affect the likelihood of successful copulation. Other sources of mating failure which may be attributed to behavioural patterns include poor management of the young boar at the onset of its reproductive life, over- and under-use of the boar, the social environment of boar and sow, and the design of the service house. It should further be added that abnormal behavioural patterns may also develop if, for example, an animal has leg or foot troubles or if a large size differential exists between male and female.

13.1 Normal patterns of behaviour

In order to establish where problems associated with mating behaviour occur, it is necessary to know what we mean by normal ethology. In this section an attempt is made to define the normal pattern of behaviour in both male and female pig at the time of mating.

THE FEMALE

The oestrous period in the sow is associated with physiological, morphological and behavioural changes (*see* Chapter 4). The latter are of particular practical importance as they allow both the stockman and the boar to detect the onset of oestrus in the sow.

The basic morphological change seen in the oestrous sow is a swelling and reddening of the vulva. As mentioned previously (Chapter 4) these changes begin 2–6 days before the onset of the oestrus. Vulval reddening and swelling usually reach a maximum before the start of the oestrous period, and are therefore subsiding during the period of sexual receptivity (Ito, Kudo and Niwa, 1960; Holtz, 1967). This is easily observed in gilts, although Holt (1959) has suggested that it occurs in only 75 per cent of sows. Other morphological changes associated with oestrus include the proliferation of vaginal leucocytes (Altmann, 1941), mucous discharge from the vulva, and a drop in vaginal pH (Holtz, 1967). Unfortunately none of these morphological changes provides an infallible method of heat detection, although vulval reddening and swelling are normally used as general indicators of the initiation of oestrus.

Behavioural patterns at the time of oestrus also tend to be variable. However, in general the sow becomes more restless as oestrus approaches, and her appetite becomes more variable. She will frequently sniff the genitals of her pen-mates, and will ride the others or allow herself to be mounted. Indeed mounting behaviour shows a definite increase in the oestrous period (*Figure 13.1*). During oestrus the sow

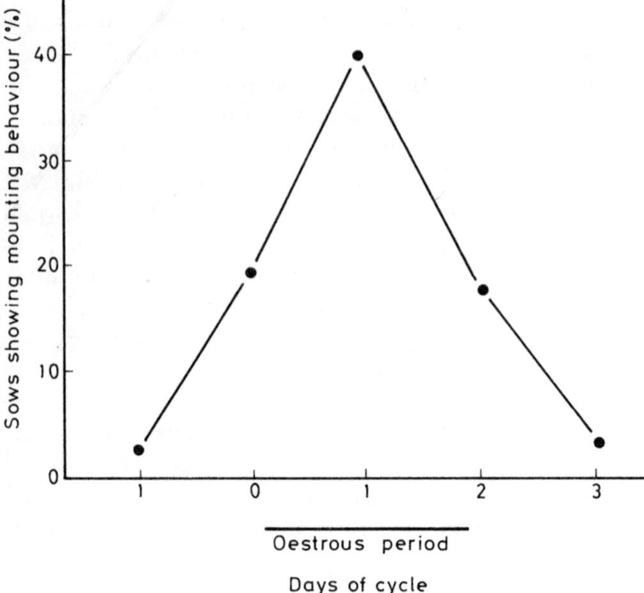

Figure 13.1 Percentage of sows displaying mounting behaviour over the oestrous period (from Holtz, 1967)

is often heard to utter grunts which are characteristic of this period, and she will increase her spontaneous activity, often to twice its normal level (Altmann, 1941). The latter activity is exploratory in nature and is directed towards identifying the location of a boar. Once this has been achieved the oestrous sow will go to the boar and remain close, even to the extent of actively pursuing him (Jakway and Sumption, 1962).

The culmination of the behavioural sequence during oestrus is the 'standing heat reflex' (lordosis), or mating stance of the sow. This is indicative of sexual receptivity, and is characterized by the sow standing absolutely immobile, arching her back, and cocking her

Table 13.2. The relative influence of boar-originated stimuli on the 'standing heat reflex' of oestrous gilts

Stimuli	*Sound, odour*	*Sound, odour, sight*	*Sound, odour, sight, contact*	*Total*
Number of 'standing heat reflexes'	681	51	25	757
Percentage of 'standing heat reflexes'	90.0	6.7	3.3	100

From Signoret (1972).

ears. This reflex is usually exhibited by the receptive sow in the presence of a boar, although it can be elicited by the application of pressure to her back (i.e. by simulating the pressure applied to the sow's back when mating occurs). This latter reaction is often referred to as 'the riding test' since the reflex may be initiated by the stockman sitting astride the sow. When the boar is present the 'standing heat reflex' in the oestrous female occurs in response to a combination of male originating stimuli. These include sound, smell, sight and contact, although the major stimuli are undoubtedly the olfactory and acoustic cues (*Table 13.2*). In the absence of a boar not all oestrous sows will exhibit the 'standing heat reflex', even in response to the 'riding test'. The data in *Figure 13.2* would suggest that at

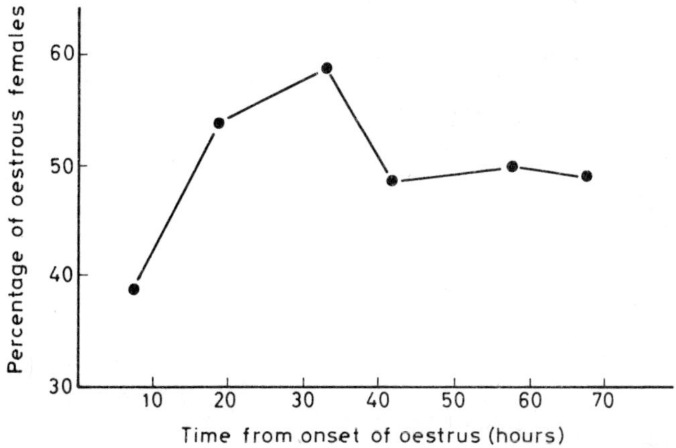

Figure 13.2 Frequency of induction of the 'standing heat reflex' by the 'riding test' at various times during oestrus (from Signoret, 1972)

most 50–60 per cent of oestrous females are likely to be detected in the absence of a boar. This, of course, has important implications when considering artificial insemination or the housing of replacement gilts and weaned sows.

THE MALE

Assuming that the boar is of normal libido he will be ready for copulation at all times. When an oestrous female is introduced to the boar he will usually mount her as soon as the 'standing heat reflex' is exhibited – indeed, it appears to be the immobility of the female that initiates mounting by the male. In some cases mounting and dismounting several times will occur before copulation begins, but this tends to be the exception rather than the rule. Most boars once mounted remain so until the end of copulation.

Following mounting, erection of the penis occurs fairly rapidly, with an average interval of 82 seconds from mounting to erection reported by Wierzbowski and Wierzchos (1968). During this phase of mating the boar has his haunches clenched together and pressed forwards, and can be seen to be thrusting. However, this thrusting tends to cease fairly rapidly, and the boar is generally immobile at the time of ejaculation (Signoret, 1972). Finally, once ejaculation has been completed (a process which normally takes 4–6 minutes) the boar dismounts.

If the female is left with the boar for the whole oestrous period he will copulate several times – an average of 7.2 copulations was reported by Signoret (1972). Furthermore, if a different oestrous female is introduced this will tend to stimulate the boar and increase the frequency of copulation. Also it is worth noting that where this occurs (i.e. the boar has more than one oestrous female present) he will often display a preference for one or more of the females or, alternatively, an aversion to certain other females.

MALE–FEMALE INTERACTIONS

It is clear that the first requirement for mating is that the male and female come into close proximity. This step is initiated by the oestrous female who actively searches out a boar during the oestrous period. The mechanism whereby the boar is detected appears to be a combination of several sensory factors. Predominant among these is the boar odour, although acoustic and visual signals also play a part (*see Table 13.2*).

In contrast, the boar does not appear to be able to distinguish between oestrous and anoestrous females. Indeed, he may actively pursue an anoestrous female, even in the presence of an oestrous female (Signoret, 1972). The boar's behaviour indicates that he considers every female he contacts as a potential candidate for copulation. He therefore initiates a courtship behaviour sequence (*see below*) with every female. The response of the female to this then determines whether or not copulation is attempted.

The initial encounter between male and female normally takes the form of naso-nasal or naso-genital contact. This is often followed by nosing of the female's flanks (a form of contact that may elicit or enhance the 'standing heat reflex') or, less frequently, by pursuit and mock fighting. The next stage in this behaviour sequence is normally attempted mounting. In the case of the oestrous female this would, of course, result in copulation. However, when mounting is attempted with an anoestrous female she will usually flee from his advances or collapse to the floor, hence prohibiting copulation. Although this pattern of behaviour is an obvious signal to the boar that the female is not sexually receptive, he will often continue with his advances for a while. Once he is satisfied that no copulation will be allowed he will move on to the next available female and begin the behavioural sequence all over again.

13.2 Abnormal patterns of behaviour

It is clear from the foregoing discussion that, allowing for minor variations, a set pattern of behavioural events occurs prior to mating in the pig. However, it is also recognized that deviations from the norm, which may profoundly influence the chances of successful mating being achieved, do occur. It is therefore considered pertinent to describe the more common behavioural abnormalities, and to discuss their probable causes. Most of these abnormalities are due to incorrect management of the boar either during its working life or, more commonly, during the rearing phase. However, some may also be due to inherent defects in the animal.

ABNORMAL ETHOLOGY CAUSED BY PHYSICAL, PHYSIOLOGICAL OR MORPHOLOGICAL DEFECTS

It is recognized that certain defects in the male and female may result in abnormal behavioural patterns at the time of mating. Leg weakness or injury is probably the most common of these defects, although in general the defective partner will attempt to follow the normal behaviour sequence. In this case it is usually the last step (i.e. mounting) which results in failure – a male with weak legs will be incapable of mounting, and a female with defective legs will be unable to withstand the weight of the mounted male.

Inflammation of the genitals (due to infection or irritation) may also result in abnormal behaviour. This situation is rare in the female, but does occur in many boars. The pain associated with copulation in these cases overrides the sex drive of the boar, and hence results in mating failure. Furthermore, if the situation is not rectified, a decrease in libido will occur and an abnormal behavioural pattern will ensue.

In the sow 'silent' oestrus (i.e. ovulation without behavioural oestrus) is reported to occur in 1.49–2.1 per cent of all oestrous cycles (Burger, 1952; Signoret, 1972). These figures are worthy of note since the occurrence of 'silent' heats is often blamed in cases where heat detection has not been carefully carried out. When a true 'silent' heat does occur the sow will reject the boar's advances, despite the fact that ovulation is taking place. A rarer variation of this abnormality is the occurrence of 'split' oestrus where oestrus behaviour is followed by a non-receptive state and then a further period of receptivity. However, this is only reported to occur in an estimated 0.18 per cent of all oestrous cycles (Burger, 1952).

A final morphological abnormality which may influence behaviour is the occurrence of intersexual females. Intersexuality without abnormal external genitalia has been reported to occur in 0.1–0.2 per cent of Large White females (Signoret, 1972). This condition usually results in the female displaying permanent sexual interest, although the boar is seldom allowed to mount.

ABNORMAL ETHOLOGY DUE TO INCORRECT MANAGEMENT

This type of abnormal behaviour may take one of four forms – aggression and fear, extended foreplay, low libido or abnormal mounting and service.

Aggression and fear

This refers to true aggression as opposed to the normal pursuit and mock fighting discussed earlier. It may be initiated by either the boar or the gilt/sow depending on the circumstances. For example, the boar will often attack a young gilt if he cannot elicit a standing heat reflex or if she repeatedly fails to support his weight when mounted. This is usually a manifestation of frustration, although it should be added that large boars do sometimes attack very small gilts as soon as they come into contact, regardless of whether or not the gilt is on heat. Frustration may also result in attacks on sows, but this is less common and is usually a result of introducing anoestrous sows to the boar.

Attacks by gilts/sows on boars are also frequently observed. Individual gilts tend to be submissive when confronted by a boar, but when presented to the boar in a group (e.g. when stimulating precocious puberty by means of boar contact) the gilts may become the aggressors. Sows, too, are often observed to attack the boar, particularly if he is relatively small. However, this may constitute part of the normal behaviour sequence since the rejection of a persistent boar by an anoestrous female may require aggression as well as the fleeing response.

The dominant aggressor usually elicits fear in the recipient and this will be expressed as a fleeing or submissive response. The latter may occasionally result in copulation with an anoestrous female, although this is fairly rare and does not, of course, result in pregnancy.

Injuries received in attacks such as those described above are normally minor, being composed mainly of small abrasions and bruising. However, economic damage is done by the failure to achieve successful mating. Also limb injuries do sometimes occur as a result of such aggression, and this may lead to subsequent mating problems.

Finally, large boars have been occasionally observed to inflict severe, and in some cases fatal, injuries on young gilts, particularly when the gilt's flanks and shoulders have been ripped upen by the boar's tusks.

Extended foreplay

This term is used to describe the situation where the courtship behaviour sequence prior to mating is excessively prolonged. It is a situation for which the boar is primarily responsible since the oestrous sow will readily exhibit a standing heat reflex in response to the male's advances. In cases where the sow is not in the peak of oestrus, however, (i.e. at the very beginning and end of the oestrous period and, of course, during the rest of the oestrous cycle) the boar may have difficulty eliciting the standing heat reflex, and extended foreplay becomes necessary. The problem arises when the sow readily adopts the rigid position of the reflex, but the boar continues with the foreplay. In these cases it usually appears that the boar has a preference for foreplay, and he will often display an unwillingness to actually mate. When this occurs the sow may suffer from the enthusiastic, though non-aggressive, attention of the boar.

In many cases of extended foreplay the sow is too exhausted to withstand the weight of the boar by the time he actually gets around to mounting. This problem occurs quite frequently, and often results in the oestrous sow not achieving successful copulation.

Low libido

The occurrence of low libido, or sex drive, is only of relevance to the boar. It usually refers to the situation where the boar shows a lack of interest in oestrous females, and is unwilling to indulge in courtship behaviour or copulation. However, it should be recognized that low libido does not refer to a set level of sex drive but rather encompasses a range of libido from a total lack of interest to an unwillingness to initiate courtship behaviour. Thus, at the extreme, the boar will totally ignore the oestrous female when introduced, and retire to a quiet corner of the pen. On the other hand, some boars having low libidos may be assisted to achieve copulation through patience and encouragement. This is quite a common problem, however, and can result once again in oestrous sows failing to achieve successful copulation.

Abnormal mounting and service

The three important variants are head mount, side mounting and anal service. All three are reasonably common, although their status as abnormal behaviour may not be fully justified.

As has been mentioned previously, the primary stimulus that initiates mounting attempts by the boar is the immobility of the oestrous female. The fact that boars will mount 'dummy sows' or other immobile boars suggests that immobility and overall shape are the main criteria by which the boar recognizes a receptive female (Signoret, 1972). Head and side mounting may therefore only be a result of this recognition procedure producing slight confusion in the boar. Furthermore, the boar frequently appears to recognize his mistake and proceeds to readjust to the correct position. In cases where the boar does not correct his own position it is usually a simple procedure for the stockman to remove the mounted boar and assist him to the correct alignment.

Anal service may also be considered to be a non-behavioural problem. The proximity of the vulva and anus does mean that the boar's penis may enter the anus by mistake at the time of copulation. The frequency of occurrence of this problem is not recorded, but it does seem likely that it is fairly common. It may, however, be avoided if all services are closely observed. In this situation the boar may be removed from the mounted position if anal service is initiated. On his second attempt the boar should achieve vulval entry or, if not, should be assisted to do so by the stockman.

13.3 Causes of abnormal behaviour

Abnormal behaviour associated with physical, physiological and morphological defects has already been discussed. Here it is intended to consider the possible causes of behavioural abnormality in otherwise normal animals. This subject will be dealt with under the headings of size differential, timing of introduction and boar management.

SIZE DIFFERENTIAL

It is clear that problems will arise at mating if the boar is considerably larger than the sow or gilt or vice versa. These are, initially, physical problems associated with the actual mechanics of achieving copulation. However, they can and do lead to behavioural problems, the most common of which is aggression. Indeed, the frustration caused by the inability to achieve copulation often causes the boar to attack the sow, and occasionally the sow to attack the boar. This has been mentioned previously in relation to the young gilt where the situation is frequently encountered. If such a problem is allowed to continue (i.e. if a large boar is frequently paired with small gilts with whom he rarely achieves copulation) it may lead to further behavioural abnormalities. First, this type of frustration could result in the boar associating small gilts with frustration, and hence result in attacks being made as soon as gilts are introduced. It is also clear that repeated failure to achieve copulation will result in adverse effects on the libido of the boar, often culminating in the problems of low libido discussed above.

TIMING OF INTRODUCTION

The observation of oestrus in the female is obviously critical in order to determine when she should be introduced to the boar for mating. The alternative is to introduce all females to the boar, since he may be considered to be a better detector of the oestrous female than the stockman. This system would, for example, result in all weaned sows being taken to the boar daily for heat detection from the day of weaning until successful mating was achieved. This, in theory, would appear to be a sensible system, since it should result in all oestrous females being detected. Unfortunately, in practice, the effect of such a system is to expose the boar to a large number of females, relatively few of which are on heat. He will initiate courtship behaviour with each female, but in this situation his success will be dependent on the proportion of oestrous/anoestrous females introduced. This may have one of two effects – either the boar shows a frustration response to his low success rate and becomes aggressive, or he recognizes that his chances of eliciting a 'standing heat reflex' in any particular introduced female are relatively small and therefore he devotes little attention to that introduced female (i.e. he suffers a drop in libido). It should, however, be emphasized that these reactions are extreme, and are unlikely to occur if the only anoestrous females introduced are those at the very beginning or end of their oestrous period which the stockman considers may show a 'standing heat reflex' in response to the boar.

BOAR MANAGEMENT

Recent experiments carried out in Australia by Hemsworth and co-workers (Hemsworth, Beilharz and Galloway, 1977; Hemsworth *et al.*, 1977) have indicated that the management of the boar in both the pre- and postpubertal periods can exert a profound influence on subsequent behaviour.

In a study of the effects of social conditions during rearing on sexual behaviour, Hemsworth, Beilharz and Galloway (1977) compared the rearing of young boars in isolation with rearing in groups (all male or mixed sex). The results, given in full in *Table 13.3*, clearly demonstrate that social restriction during rearing causes a severe reduction in the boar's courtship behaviour and copulatory performance. Indeed, the authors concluded that boars not exposed to visual

Table 13.3. The effects of social conditions during the rearing period on sexual behaviour in boars

	Mean of rearing group		
	Social restriction	*All male group*	*Mixed sex group*
Courtship behaviour			
Number of 'nosing the sides of gilts'	28.4	128.2	83.5
Number of 'chants'	30.6	73.0	121.0
Number of 'naso-nasal contacts'	102.0	118.0	145.2
Number of 'naso-genital contacts'	58.6	60.9	126.5
Number of mounts	35.4	112.9	135.4
Sum of all courtship behaviour activities	320.0	572.0	690.6
Copulatory performance			
Number of copulations	2.1	10.6	8.5
Time spent ejaculating (min)	5.8	36.9	31.1
Average duration of ejaculation (min)	2.1	3.6	3.8
Average reaction time to 1st mount (min)	4.7	1.4	1.3
Percentage of mounts that were head mounts	6.5	2.0	1.8

From Hemsworth, Beilharz and Galloway (1977).

or physical contact with other pigs during the rearing period were of low sexual motivation and exhibited an impaired ability to react socially to other pigs. Those boars that were reared in all male groups also tended to exhibit less frequent courtship behaviour than did boars reared in mixed sex groups. However, since this result was not reflected in copulatory performance, it cannot be considered to be a problem. Thus, it would seem that the stimuli provided by the presence of prepubertal gilts or prepubertal boars during rearing may make a similar contribution to the sexual behaviour of the mature boar. Finally, the results of Hemsworth, Beilharz and Galloway (1977), together with results obtained by other workers (Zimbaldo, 1958; Folman and Drori, 1965; Gerall, Ward and Gerall, 1967), indicate that the low level of copulatory performance exhibited by boars reared in isolation is of a permanent nature, and cannot be reversed by allowing frequent opportunities for mating following the attainment of maturity.

The effects of social conditions on sexual behaviour in the mature boar are equally severe, although they do appear to be reversible (Hemsworth *et al.*, 1977). Boars housed near to receptive females display significantly higher levels of courtship behaviour and copulatory performance than boars housed in isolation or adjacent to other boars (*Table 13.4*). These results differ from those obtained in the prepubertal boar in two respects. First, those boars housed in association with other males exhibited similar behaviour patterns to boars kept in isolation – that is, the presence of sexually receptive females is necessary to maintain high levels of sexual behaviour in the mature boar. Also, when the socially restricted boars (i.e. those housed in isolation or adjacent to other boars) were placed in pens adjacent to receptive females their sexual behaviour was restored to normal within 4 weeks (*Table 13.4*: period 2). Thus, it is clear that

Table 13.4. The effects of social conditions on sexual behaviour in mature boars

	Mean of group		
	Social restriction	*Housed near mature boars*	*Housed near receptive females*
Period 1			
Sum of all courtship behaviour activities	167.8	192.5	323.8
Number of copulations	1.25	1.50	7.75
Time spent ejaculating (min)	2.63	4.25	21.75
Period 2 (all boars housed near sexually receptive females)			
Sum of all courtship behaviour activities	280.0	291.5	248.0
Number of copulations	3.00	3.00	3.75
Time spent ejaculating (min)	8.50	9.50	11.38

From Hemsworth *et al.* (1977).

the problems associated with housing boars in the absence of receptive females are not permanent but, rather, may be reversed within a short period of exposure to oestrous females.

13.4 Conclusions

The overall effect of a failure at mating is that the sow remains unproductive for another 21 days (i.e. one oestrous cycle). This is precisely the same result as that obtained when failure occurs at either the conception or fertilization stage, and yet these latter periods are viewed with much greater concern than are the behavioural problems associated with mating. Indeed, it may be speculated that many of the returns to service attributed to conception failure are, in fact, due to failure to achieve true copulation. If this proves to be the case then the causes and cures of abnormal behaviour, together with the methods available to prevent them, are of considerable importance.

The first stage at which mating behaviour may be influenced is during the prepubertal period. The results of Hemsworth and co-workers have demonstrated that social conditions at this time may exert a profound and irreversible influence on the subsequent sexual behaviour of the mature boar. The correct form of rearing for young boars would appear to be in social groups (either all male or mixed), since social restriction, such as that imposed by rearing in isolation, is the major cause of abnormal behaviour attributable to this period. Such conditions may be met in practice by allowing young boars to have contact with other pigs, these being housed in either the same pen as the boar or a pen adjacent to him. In addition, the young boar should be provided with clean, warm, dry accommodation, and fed in such a way that he remains in good condition without becoming fat (this may be achieved in practice by feeding to an incremental scale up to a maximum of 2.7 kg feed/day).

The end of the rearing period may be considered to occur when the boar achieves his first mating. This is a crucial point in a young boar's life and may, indeed, determine whether he is maintained in the herd or culled for poor performance. It is at this time that the boar may attain confidence or suffer frustration, depending on whether or not his first service is successful. Since this one service is so important it is worthwhile taking every available step to ensure its success. First, a small docile sow which is firmly on heat should be chosen as his first mate – gilts should be avoided as they tend to become excitable in the presence of a boar. Next, the service itself should be supervised from start to finish. Assistance should be given to ensure that the boar attains the correct mounted position and achieves vaginal insertion of the penis. Once this has occurred success is virtually assured. Following such a successful first service the young boar's confidence will increase rapidly, and the chances of subsequent failure are minimized. However, it should be emphasized that over-use in this initial period is not advisable. One 'double' service a week is adequate over the first few weeks of use, this being increased to a mature boar's workload of three two services per week by 12 months of age.

Once the boar is mature and serving normally the aim must be to maintain his fertility and libido. The first priority is to provide him with an optimum number of services, since under-use can result in as many problems of fertility and libido as over-use. Approximately three 'double' services per week should be adequate for this purpose, although this may be stepped up for short periods without adverse effects. When feeding the mature boar the objective is to keep him 'fit not fat', and therefore feed levels should be varied in order to meet this requirement (usually in the range 2.7–3.6 kg feed/day). The boar should be housed in close proximity to sows or gilts since the evidence suggests that boars kept in isolation or only in contact with other males display reduced copulatory performance. An ideal way of meeting these requirements would appear to be a service house in which individual boar pens alternate with group pens housing either weaned sows or replacement gilts. This arrangement has two advantages: first it ensures that social conditions do not adversely influence the performance of the boar, and second the presence of the boar will provide a stimulatory influence on puberty attainment in the gilts and the manifestation of the postweaning oestrus in the sows. This form of service house design has the further practical advantage that the movement of gilts and sows or boars at the time of service is minimal.

When the gilt or sow is taken to the boar for mating the first priority is to ensure that she is actually on heat. Previous discussion has highlighted the possible adverse effects that persistent exposure to anoestrous females may have on the boar. However, it is clearly not always practical to identify sows that are on heat. It is therefore recommended that sows which are definitely anoestrous are not taken to the boar and thus he is only exposed to those females suspected of being, or identified as being, on heat. Such a system should not result in any marked decline in the fertility or libido of the boar.

One other factor worthy of reiteration is the supervision of services. It was emphasized earlier that this was of vital importance when the young boar was experiencing mating for the first time. However, it is suggested that every service should, in fact, be supervised. By so doing abnormalities in either courtship or copulatory behaviour may be quickly recognized and remedial action taken. Furthermore, the occurrence of anal service may be observed and corrected. In this way the number of returns to service due to mating abnormalities may be reduced to a minimum.

In conclusion, therefore, it may be stated that there is a considerable number of possible abnormalities and problems that may occur at mating. The presence of these abnormalities can, and undoubtedly does, have an adverse influence on the overall reproductive efficiency of a herd. However, since they may be readily avoided by correct management, it is suggested that the efficiency of many herds could be improved by the adoption of the simple procedures outlined above.

13.5 References

ALTMANN, M. (1941). *J. comp. Psychol.* **31**, 481
BURGER, J.F. (1952). *Onderstepoort J. vet. Res.* **2**, 3
FOLMAN, Y. and DRORI, D. (1965). *Anim. Behav.* **13**, 427
GERALL, H.D., WARD, I.L. and GERALL, A.A. (1967). *Anim. Behav.* **15**, 54

HAFEZ, E.S.E., SUMPTION, L.J. and JAKWAY, J.S. (1962). In *The Behaviour of Domestic Animals* (Ed. by E.S.E. Hafez), pp. 370–396. London, Ballière, Tindall and Cox

HEMSWORTH, P.H., BEILHARZ, R.G. and GALLOWAY, D.B. (1977). *Anim. Prod.* **24**, 245

HEMSWORTH, P.H., WINFIELD, C.G., BEILHARZ, R.G. and GALLOWAY, D.B. (1977). *Anim. Prod.* **25**, 305

HOLT, A.F. (1959). *Vet. Rec.* **71**, 184

HOLTZ, W.H. (1967). *Studien zum zyklischen Ablauf der Sexualfunktionen und zer Methodik der Zyklusdiagnose bei weiblichen Zwergschweine.* Gottingen, Diss

HUGHES, P.E. and COLE, D.J.A. (1978). *Anim. Prod.* **27**, 11

ITO, S., KUDO, A. and NIWA, T. (1960). *Natn. Inst. Agric. Chiba-Shi, Japan,* 27

SIGNORET, J.P. (1972). In *Pig Production* (Ed. by D.J.A. Cole), pp. 295–314. London, Butterworths

WIERZBOWSKI, S. and WIERZCHOS, E. (1968). 6th Congr. int. Anim. Reprod. A.I., Paris, II, 1681

ZIMBALDO, P.G. (1958). *J. comp. physiol. Psychol.* **51**, 764

Chapter 14

Artificial insemination

Artificial insemination in pigs was first carried out in the 1930s in the USA, USSR and the Philippines but the first commercial use of the technique in the UK appeared about 1955. Since this time we have seen an exponential increase in the use by dairy farmers of artificial insemination but over the same period little increase in pig A.I. There is a multiplicity of factors responsible for this disparity which will be discussed in the present chapter.

In 1965 however, the Semen Delivery Service was introduced by the Pig Industry Development Authority (now the Meat and Livestock Commission) and A.I. is available to any pig producer in the UK through this and other similar schemes operated by commercial organizations.

It is difficult to assess the exact usage of A.I. relative to natural service but the figure is probably somewhere between 4.5 and 5.0 per cent. This assumes about 62 000 inseminations are carried out artificially every year in the UK.

Undoubtedly this figure will increase when a reliable long term storage method has been developed but progress with conception rate is needed before A.I. will become very widely employed.

Livestock improvement is the biggest single advantage to accrue from the use of A.I. in pigs. Only the top 5 per cent of Meat and Livestock Commission (M.L.C.) performance tested boars are eligible for entry to the A.I. stud and in practice the boars kept at the A.I. stations are at the pinnacle of quality. An analysis of some 5148 slaughter pigs has shown (M.L.C., 1975a) that each of the progeny of A.I. boars was worth on average about 5 per cent more to the producer than carcasses sired by natural service. This was due to the higher lean content, decreased fat level and improved feed efficiency. Hence the profitability of a 200 sow bacon unit could be considerably elevated if A.I. was used in the whole herd. This of course assumes none of the disadvantages which will be discussed later.

Other benefits which are to be gained from the use of A.I. are: reduced cost of service as the need to purchase, feed and house a boar is obviated, important particularly for the smaller herd; reduced disease risk, due to introduction of new boars to the herd; and finally the risk to stockmen is removed with no entire males on the unit. The latter point is probably a minor consideration but is relevant again to small part-time units. *Table 14.1* gives the results from a survey of A.I. users and shows the proportion of them using the service for the various reasons given. Clearly the majority of users exploit the availability of high genetic-merit sires at low cost.

Table 14.1.

Reason used	*Percentage of user producers*
Crossbreeding programme	13
Batch farrowing	19
Genetic improvement	78
Reduced disease risk	10
Reduced cost	14
Miscellaneous	13

From M.L.C. (1975).

Batch farrowing units may benefit from A.I. Where farrowing is in large groups twice a year, for example, the acquisition of a large number of boars for just 2 weeks a year would be uneconomic. Crossbreeding is also given as a reason by A.I. users. These producers tend to be pedigree breeders who are producing crossbred gilts for sale.

14.1 Collection and storage of semen

There is little doubt that boars are easily trained to mount simple dummy sow devices and will ejaculate readily without sophisticated artificial vagina devices. Chamohoy, Abilay and Paled (1960) found that it took 7 days for inexperienced boars to be trained to mount a dummy and ejaculate freely. Boars which had previously experienced some sexual activity (natural service) took an average of 14 days before working to the satisfaction of the operator. Niwa (1961) has recommended that training for collection purposes should start at 100 kg liveweight but this may be difficult to put into practice owing to the fact that a boar entering an A.I. stud is often heavier than this at entry.

The frequency of collection is also an important consideration in relation to the average yield of spermatozoa. Gerritts, Graham and Cole (1962) found that over a 20-day period 5, 10 and 20 ejaculations yielded an average total yield of spermatozoa of 54.9×10^9,

39.9×10^9 and 23.7×10^9 respectively. Niwa (1961) thus recommended a 5–6 day collection interval which is in practice adhered to at M.L.C. stations.

Collection is carried out following the mounting of the dummy by the boar. The boar's prepuce is massaged to evacuate as much preputial fluid as possible and then with the operator wearing a clean surgical glove, pressure is applied to the boar's erect penis until ejaculation is complete. The first fraction (low in sperm) is discarded and the second fraction which is sperm rich is collected in a warmed (30 °C) reagent bottle. It is filtered using a milk filter and plastic funnel.

Immediately after collection it is important to ensure temperature shock does not occur and the reagent bottle in which the semen is collected should be contained within a vacuum flask to make sure no rapid change in temperature occurs. The semen is then transferred to a measuring cylinder kept in a constant water temperature bath at 28 °C.

At this stage the semen is evaluated for its fertility characteristics. A drop is placed on a heated microscope stage at 35 °C and given a subjective rating depending on the vigour of wave motion (seen under low power) and of flagella motion (under high power). Sperm density is then estimated by the use of a colorimeter previously calibrated for the optical density of different sperm solutions of known concentration.

The appropriate amount of diluent/extender is now added to give a sperm concentration of constant amount ready for despatch. Individual doses are transferred to 50 ml plastic bottles and stored at 15–20 °C until they leave the station in insulated containers.

The appropriate amount of diluent is then added to leave a constant sperm concentration in each millilitre of semen. This means in practice that each ejaculation will provide on average 25 double doses.

A diluent which has been used successfully in recent years is the so-called Guelph extender. *Table 14.2* gives the constituents of this extender which are added to distilled water to make up 1 litre of the diluent.

Table 14.2. Composition of the Guelph diluent/extender

Component	*Weight per litre*
Glucose	60 g
Trisodium citrate	3.7 g
Ethylenediaminetetra acetic acid	3.7 g
Sodium hydrogen carbonate	1.2 g
Penicillin	500 000 i.u.
Streptomycin	500 mg

This diluent gives a dose of semen a viable 'shelf' life of between 5 and 7 days at room temperature. Another extender which has probably been most widely used is the Illinois Variable Temperite (IVT) diluent (Du Mensil du Buisson, Jondet and Locattelli, 1961) which gives a storage life of 3–5 days. This contains glucose, sodium hydrogen carbonate, sulphanilamide, sodium citrate and potassium chloride together with antibiotics. Carbon dioxide is also bubbled through the mixture.

The viability of undiluted semen is between 2 and 24 hours after ejaculation and this is extended by the addition of these diluents. Very recently a commercial organization in the UK has developed a long-life extender known as SCK-7 (Kerekgyarto, 1977) which it is claimed allows a fertile insemination up to 6 days following ejaculation. If this claim is upheld then this could represent increasing flexibility for operators of A.I. schemes and potentially higher farrowing rates. Results have shown that from 2949 litters the percentage of sows farrowing to the first insemination was 79.1 per cent. It is hoped also in the near future to make available an extender to prolong the viable life of semen doses to 10 days.

Early attempts to use deep-freezing techniques for the long term storage of semen were generally unsuccessful (Einarsson, 1973) but Polge, Salamon and Wilmut (1970) demonstrated that fertilizing capacity for sperm cells following freezing and rethawing was maintained at least in part. The advantages to be gained from such a technique are many:

(1) To facilitate import and export of semen.
(2) To stockpile semen of high genetic merit.
(3) To monitor national genetic improvement more efficiently.
(4) Allow nominated semen even following the death of the sire.
(5) Potentially to allow users to hold their own stocks of semen for more efficient timing of insemination.

Some measure of success has now been achieved with freezing techniques and this topic has recently been concisely reviewed by Larsson (1979). Most workers favour some form of sperm concentration method by centrifugation. Cryoprotectants are also currently being developed which give far better sperm survival and subsequent fertility than previously experienced.

Polge *et al.* (1970) showed that the main problem with the freezing of boar semen was that the motility and transport of sperm cells were affected and by surgically implanting semen into the oviducts normal fertility resulted.

Larsson (1979) describes four methods for the processing and freezing of semen and *Table 14.3* gives the pregnancy rates and litter sizes using these methods.

Table 14.3. Pregnancy rates and litter sizes after insemination of boar spermatozoa frozen according to the methods of the given authors

Method	*Reference*	*Number of sows/gilts*			*Mean litter size*
		Inseminated	*Pregnant*	*%*	
Pursel and Johnson	Johnson *et al.* (1978)	33	27	81.8	9.5*
		144	92	63.9	–
Westendorf *et al.*	Richter *et al.* (1976)	201	139	69.2	8.4
Paquignon and Courot	Paquignon *et al.* (1977)	138	80	57.8	9.7
Larsson *et al.*	Larsson *et al.* (1977)	36	28	72	9.7*

* Estimated at slaughter 28 days post coitum.

The results from *Table 14.3* indicate that it is now possible to achieve a reasonable degree of success with deep freezing techniques but as Larsson (1979) points out, 'The methods are sperm- and labour-consuming and should so far be regarded as complementary to utilization of liquid semen (fresh) for routine pig A.I.' In the United States and Canada however there is now a commercial service using the method of Pursel and Johnson. This method takes the sperm rich fraction of the ejaculate and after adding antibiotics the semen is allowed to cool to room temperature over a 2 hour period.

Table 14.4. Composition of diluents (method of Pursel and Johnson)

Ingredient	*Amount*
BF5	
TES-N-TRIS ((Hydroxymethyl) methyl 2 aminoethane sulphonic acid)	1.2 g
TRIS ((Hydroxymethyl) amino methane)	0.2 g
Dextrose, anhydrous	3.2 g
Egg yolk	20 ml
Orvus Es Paste	0.5 ml
Distilled water	100 ml
BTS	
Glucose	4.0 g
Sodium citrate dihydrate	0.6 g
Sodium bicarbonate	0.125 g
EDTA	0.125 g
Potassium chloride	0.075 g
Distilled water	100 ml

The semen is then dispensed in aliquots containing 6×10^9 spermatozoa in centrifuge tubes. Centrifugation is performed at 300 *g* for 10 min. The seminal plasma is removed and the spermatozoa are resuspended to a total volume of 5 ml per tube with BF5 extender (*see Table 14.4*). The extended semen is cooled to 8 °C over a 2.5 hour period. After this, 5 ml of BF5 containing 2 per cent glycerol is added to each tube. The contents of each tube are immediately mixed and frozen in 0.2 ml pellets on dry ice.

For thawing, 10 ml of pellets (equivalent to one insemination dose or 6×10^9 spermatozoa) is transferred to a dry thawing box and kept there for 3 minutes before being placed into 45 ml of 50 °C BTS thawing diluent. Insemination can then be carried out.

It is now certain that different boars will 'freeze' better than others (Larsson and Einarsson, 1976; Polge, 1976; Paquignon *et al.*, 1977) and this is due to genetic differences between boars associated with a varying resistance of sperm cells to processing, freezing and thawing. In view of this effect it is necessary to screen boars to be used for A.I. for freezing ability. One problem is that a suitable assay technique to estimate freezing ability quickly is needed. At present effective assays have only been developed on a laboratory scale although a number of different assays are under investigation (Larsson, 1979).

There are therefore signs that progress is being made with freezing of boar's semen but at present the vast majority of A.I. schemes rely on a liquid semen service and the logistics of semen movement is probably the most important factor in such schemes.

14.2 A.I. in practice

The main problem with our current state of A.I. technology is the transportation of semen to the producer before viability is lost.

Most if not all systems in the UK rely on a 'Postal Delivery Service': semen doses are despatched on the day of collection by the national postal service and should arrive on the farm within 24 hours. This is backed up by a rail despatch service which is quicker by about half a day. Speed is of obvious importance as a producer may see a sow or gilt on heat in the early morning hours and if the semen arrives after 24 hours then the optimum insemination time has passed and fertility may be much reduced.

When the semen arrives on the farm one of the doses should be used immediately and the second dose given 12–16 hours later. The actual insemination involves the insertion of a specially designed rubber catheter into the sow's vagina until the spiral tip is 'locked' into the cervix. When the lock has been established the bottle containing the semen dose is attached to the end of the catheter and the semen runs in under gravity. This usually takes between 7 and 12 minutes but may last as long as 20 minutes. It is important to exercise patience at this time as any forcing of semen in by squeezing the plastic bottle will result in a backflow and leakage. When all the semen has run in the catheter is carefully withdrawn and sterilized in boiling water before the next insemination.

14.3 Problems with A.I.

In view of the fact that access to the top national sires is so cheap and easy one might have expected that more producers would use the services available to upgrade their herds in the long term and improve their profitability in the short term.

One of the major problems is that producers feel that the inconvenience of ordering semen, carrying out the insemination and looking after equipment cancels out any benefits. Certainly time is required to carry out the routine properly but even with natural service there is a necessity to spend quite a few man-hours each day in heat detection and supervision of service if good results are to be achieved. The insemination technique also is straightforward with no particular skills required over and above those which should be possessed by good stockmen.

Another reason why A.I. is not so widespread is that average herd size has grown significantly in recent years and as a result the average economics of keeping one's own boar(s) are far more favourable. The largest herds can in fact purchase good quality genetic material themselves and justify the expense easily on the grounds of the number of progeny reared per year or again as part of an upgrading policy.

Certainly the biggest single drawback to more widespread application of A.I. in pigs is the fact that return rates are higher than for natural service and producers are well aware of this. In actual practice there is a tremendous variability attached to this observation and there may be many different reasons why it should occur at all.

Most reports indicate that conception rate drops by somewhere between 5 and 15 per cent below that expected with natural service. There are some indications that the farmer-operated semen delivery service maintains better results than A.I. technicians who are centrally based and travel out to perform inseminations for farmers.

At first sight this seems an unexpected observation but in fact the farmer doing the insemination himself has a better chance of getting the timing right than the inseminator on his rounds and it is in no way a reflection of the competence of the technician. Konnermann (1974) has observed this effect (*Table 14.5*).

Poor detection of oestrus in the absence of a boar may be a large contributory factor to poor farrowing rates as checking for returns

Table 14.5. One versus two inseminations in the same oestrus

Operator	*Conception rate (%)* *Number of inseminations* *One*	*Two*
A.I. technician	67.0	67.85
Farmer	75.4	65.25

From Konnermann (1974).

is that much more difficult. In addition difficulty is experienced by some producers in detecting the first postweaning oestrus and hence sows may eventually be served with a much extended time from weaning to conception. Holt (1959) has observed that only 75 per cent of sows exhibited vulval changes at the time of oestrus and only 60 per cent showed the 'standing reflex' when artificially induced by the herdsman sitting astride the sow's back (Signoret and Du Mesnil du Buisson, 1961). Teaser boars are hence desirable to increase the detection rate. Signoret and his colleagues have in fact identified three types of stimuli, all of which must be present if a good response is to be elicited: auditory, olfactory and tactile. As an aid to heat detection there is now available an aerosol spray impregnated with a steroid compound with pheromonal properties which can in some instances be effective in bringing out oestrous symptoms in problem sows. In trials with static pheromonal emitters placed in the lying area of weaned sows a poor response was obtained and the pheromonal steroid did not improve the heat detection rate (Varley, unpublished data).

Incorrect timing of insemination is probably the biggest single factor responsible for poor conception rates. We have seen in previous chapters that ovulation occurs during the second half of oestrus with some variability and hence the optimum time for insemination is a few hours prior to this. Heat detection should therefore be carried out twice a day if possible and even with this executed thoroughly it can never be guaranteed that the first observed heat sign is the beginning of oestrus. Double insemination therefore gives a greater chance of conception compared to a single insemination. Swensson (1977) and Meding and Rasbech (1968) have investigated the proportion of sows conceiving to the first and to the second insemination and the results show clearly the need to double inseminate.

Table 14.5 illustrates the results of an experiment using Landrace semen for the first insemination and Large White semen for the second. By simply looking at the offspring it was then a simple matter to determine which insemination had been the effective one. It appears that the initial insemination is the more effective but without the second a considerable number of sows would not be

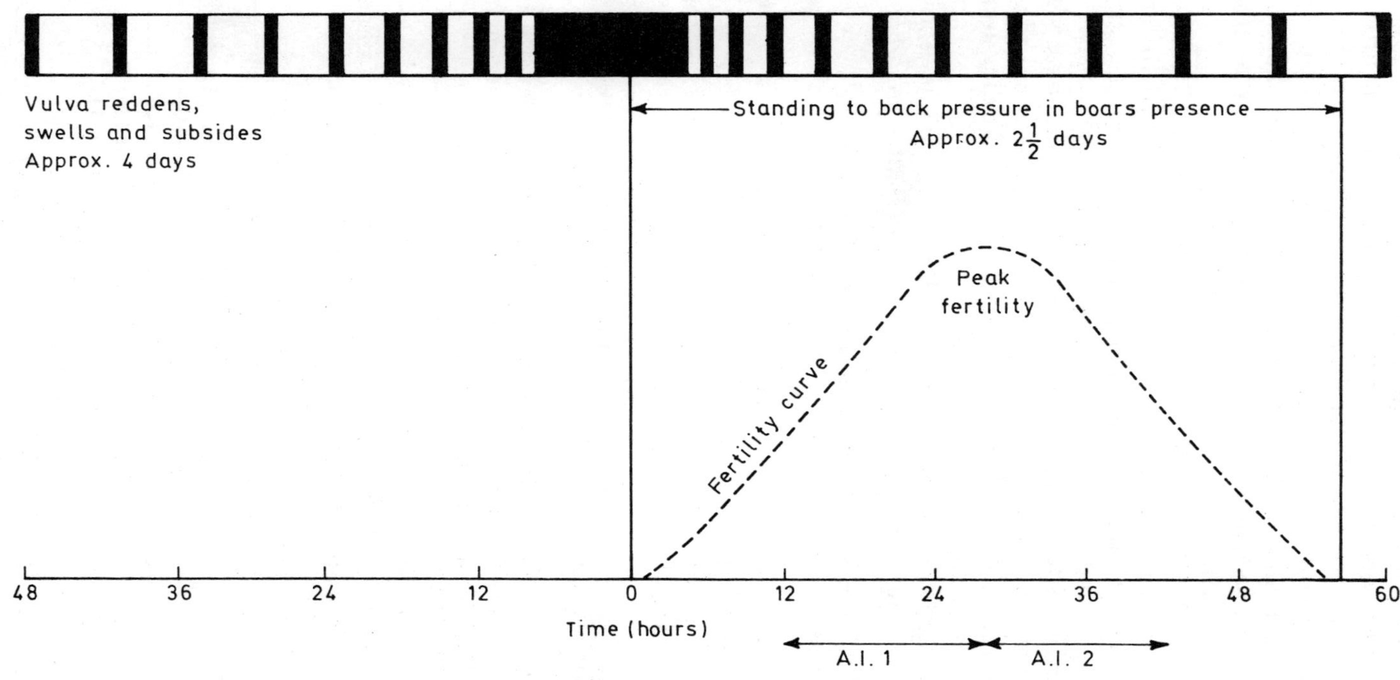

Figure 14.1 The time for insemination

effectively inseminated. It is interesting to note also that sows conceiving to the second insemination also have a reduced litter size, perhaps owing to a reduced fertilization rate arising from aged eggs.

Figure 14.1 outlines the M.L.C.'s general recommendations for timing of insemination. An interval of between 8 and 16 hours from the first to the second insemination is currently suggested as offering the best fertility level.

The size of the litter born at the next farrowing is often given as a reason for not using A.I. There is in the literature much difference in observations relating to this point. Madden (1959) in a comparative study found that A.I. litters were significantly smaller than natural service litters. Aamdal (1964) has made a similar observation: the A.I. litters from gilts averaged 1.04 fewer piglets and the A.I. litters from sows had 0.68 piglets less than the corresponding averages from sows naturally mated.

A later study by Moretti, Sassi and Dallari (1975) has shown a much smaller difference and the results are presented in *Table 14.6.*

Table 14.6. Comparison of A.I. versus natural service for prolificacy

	Number of litters	*Non-return (%)*	*Litter size*
Sows natural service	253	87.7	9.7
Sows A.I.	284	88.7	9.4
Gilts natural service	32	78.1	8.6
Gilts A.I.	102	80.3	8.4

From Moretti, Sassi and Dallari (1975).

Skjervold (1975) has also observed a difference in litter size and it can be seen from his results in *Table 14.7* that the difference increases from about 0.3 of a piglet for the first parity to 0.7 of a piglet by the third parity.

It is most likely that these differences derive from differences in the effectiveness of fertilization due to incorrect timing of insemination and also possibly to increased embryo mortality (Alanko, 1976).

Table 14.7.

Parity number	*Natural service*		*A.I.*	
	Number of litters	*Litter size*	*Number of litters*	*litter size*
1	22 727	9.56	6425	9.23
2	18 157	10.7	3511	10.24
3	11 778	11.43	2367	10.78
4+	17 550	11.57	3284	10.93

From Skjervold (1975).

McGloughlin (1976) has in fact observed no difference between natural service and A.I. for litter size and it seems probable that on some breeding units where timing of insemination is very precise, giving low return rates, litter size is not affected to a detectable degree. This hypothesis is supported by the observation of Nielson (1976) who has demonstrated a significant correlation between conception rate and litter size such that average litter size was increased by 0.12 piglets for every 1 per cent increase in conception rate.

14.4 Future possibilities and conclusions

A major step forward for the use of A.I. in pigs will be made when the technique of deep freezing of semen is perfected and as yet this development has not occurred. Currently, however, there are one or two projects under way which might prove of significance to A.I. techniques. One of these is the 'Wallsmeta', an electronic device developed by one of the commercial pig breeding companies to measure the electrical resistance of the vaginal mucosa. This electrical resistance alters over the course of the period of oestrus in the gilt and sow and it is claimed that the time of ovulation can be predicted with precision. Timing of insemination therefore becomes much easier and as only one dose of semen is required instead of two a greater economy in usage of available semen should be gained. The success of such devices hinges on the accuracy of calibration and also on the effectiveness of vaginal electrical resistance as a predictor of ovulation. As yet the nature of these relationships has not been fully determined.

At present the use of A.I. is restricted because of poor performance in terms of conception rate. Some of the variation in this is certainly associated with differences in user technique and the M.L.C. have observed that the producers who are using the most doses in a year achieve the best conception rates (*Table 14.8*). It would appear therefore that producers who use the service most become more skilled with the technique and as a consequence get better results.

Table 14.8. Conception rate and its association with the number of doses per year

Doses used per year/producer	*Conception rate* (%)
⩾ 500	80
100–499	75
20–99	73
6–19	72
⩽ 5	65

From M.L.C. (1976b).

It is probable that producers with lower conception rates tend to detect heat incorrectly and inseminate at the wrong time (s), although this has never been proved one way or the other.

It has been observed that producers who have difficulty in maintaining good conception rates also have the most problems with leakage or backflow of semen from the vagina and also with inducing a 'lock'.

In conclusion it seems unlikely that there will be an upsurge of A.I. usage within the next few years but there are significant advantages to be gained for most herds by using the technique. In particular the herd striving to upgrade its genetic merit can do so quickly and cheaply by first selecting especially good sows or purchasing 'grandparent gilts'. These are put to A.I. boars and the progeny are used as a nucleus from which to produce the next generation of replacements. This kind of breeding programme significantly reduces the costs of buying in hybrid gilts while still maintaining a high quality herd. When the time comes to replace the original 'grandparents' it is simply a matter of putting existing grandparent sows to the same breed using A.I. semen to give another generation of nucleus stock.

In this way it is possible to have an ongoing and progressive programme while still keeping the herd almost totally closed which considerably reduces disease risk. The bulk of the slaughter generation is still bred by natural service and hence there should be no major problems with litter size and conception rate. The system therefore exploits the considerable genetic advantages from A.I. and at the same time avoids the commercial disadvantages.

14.5 References

AAMDAL, J. (1964). *Proc. 5th int. Congr. Anim. Reprod., Trento* **4**, 147

ALANKO, M. (1976). *Anim. Breed. Abstr.* **44**, 3820

CHAMOHOY, L.L., ABILAY, J.P. and PALED, D.A. (1960). *Vet. Rec.* **78(5)**, 159–167

DU MESNIL DU BUISSON, F., JONDET, R. and LOCATTELLI, A. (1961). *Vet. Rec.* **78(5)**, 159–167

EINARSSON, S. (1973). *Wld Rev. Anim. Prod.* **IX**, 45

GERRITTS, R.J., GRAHAM, E.F. and COLE, R.J.A. (1962). *J. Anim. Sci.* **21**, 1022

HOLT, A.F. (1959). *Vet. Rec.* **71**, 184

JOHNSON, L.A., AALBERG, J.G., WILLEMS, C.M.T., SYBESMA, W. (1978). *Proc. 5th I.P.V.S. Congr., Zagreb*, p. KB8

KEREKGYARTO, S. (1977). *Pig Fmg* **25**, 26–29

KONNERMANN, H. (1974). *Anim. Breed. Abstr.* **41**, 142

LARSSON, K. (1979). *Wld Rev. Anim. Prod.* **XIV**, 59

LARSSON, K. and EINARSSON, S. (1976). *Acta vet. scand.* **17**, 43

LARSSON, K., EINARSSON, S. and SWENSSON, T. (1977). *Acta vet. scand.* **17**, 83

MADDEN, D.H.L. (1959). *Vet. Rec.* **71**, 227
McGLOUGHLIN, P. (1976). *Anim. Breed. Abstr.* **44**, 550
MEAT and LIVESTOCK COMMISSION (M.L.C.). (1975a). *Newsletter Pig A.I. 1975*
MEAT and LIVESTOCK COMMISSION (M.L.C.). (1975b). Pig Breeding Centres. *Analysis of herd data for 1973–1975*
MEDING, J.H. and RASBECH, N.D. (1968). *Anim. Breed. Abstr.* **36**, 379
MORETTI, M., SASSI, A. and DALLARI, L. (1975). *Anim. Breed. Abstr.* **44**, 318
NIELSON, H.E. (1976). *Anim. Breed. Abstr.* **44**, 1753
NIWA, T. (1961). *Vet. Rec.* **78(5)**, 159–167
PAQUIGNON, M. and COUROT, M. (1976). *Proc. 8th int. Congr. Anim. Reprod., Cracow* **4**, 1061
PAQUIGNON, M., BUSSIERE, J., BARITEAU, F. and COUROT, M. (1977). *Journées Rech. Porcine en France* 1977, pp. 19–21. Paris, L'Institut Technique du Porc
POLGE, C., SALAMON, S. and WILMUT, I. (1970). *Vet. Rec.* **87**, 424
POLGE, C. (1976). *Proc. 8th int. Congr. Anim. Reprod., Cracow* **4**, 1061
RICHTER, L., WESTENDORF, T. and TREU, H. (1976). *Proc. 4th I.P.V.S. Congr., Ames*, p. D16. Iowa
SIGNORET, J.P. and DU MESNIL DU BUISSON, F. (1961). In *Pig Production* (Ed. by D.J.A. Cole). London, Butterworths
SKJERVOLD, H. (1975). *Zectschrift Furtierzuchtung und Zuchtungsbiologic* **92(4)**, 252
SWENSSON, T. (1977). *Anim. Breed. Abstr.* **45**, 196
WESTENDORF, F.P., RICHTER, L. and TREU, H. (1976). *Dtsch. tieraztl. Wschr.* **82**, 261

Chapter 15

General conclusions and practical recommendations

The objective of this book has been to review each part of the reproductive life of both male and female pig, and to consider how maximum efficiency may be attained in each. To this end each chapter has dealt with a specific aspect of reproduction with a view to:

(1) explaining the natural sequence of events that occur
(2) discussing those factors which influence these events and
(3) considering how this information may be put to use in the commercial situation to improve overall reproductive efficiency.

It is now intended to summarize the improvements that application of this knowledge may make to productivity in the breeding herd. Clearly, there is plenty of room for improvement. *Table 1.1* (p. 2) shows that the average annual sow productivity for eight of the major pig producing nations is only 11.7 pigs sold/sow/year. Since a litter size at birth of 11–12 piglets has been suggested as a target (Chapter 8) and the sow is capable of producing at least two litters per year, even on 8-week weaning systems, this would suggest that the potential exists to improve productivity by up to 100 per cent. How, then, can this improvement be made?

Figure 15.1 gives a summary of the breeding lives of both male and female pigs from birth through to culling. Within this framework we can identify three specific areas of reproduction. These are:

(1) Management of the gilt from birth up to and including first mating.
(2) Productivity in the breeding sow.
(3) Boar management for optimum fertility and libido.

Once maximum efficiency has been attained in each of these areas of reproduction then the productivity of the breeding herd and hence profitability should be at a maximum.

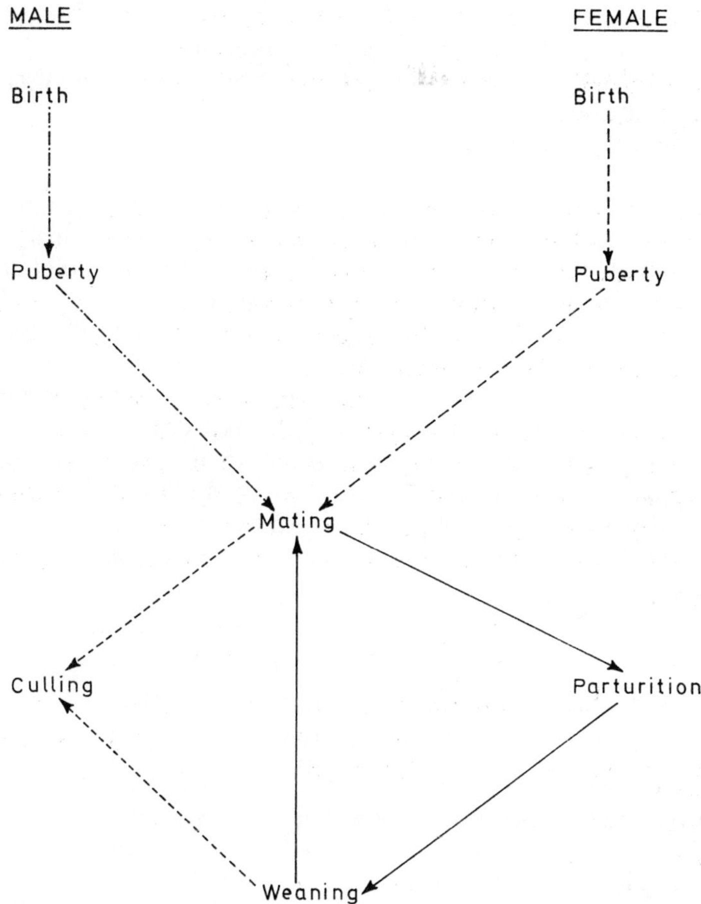

Figure 15.1 A diagrammatic representation of the breeding lives of male and female pigs

15.1 The gilt

In the average herd gilts will account for between 15 and 30 per cent of all breeding females. This being the case it must be accepted that the replacement gilt will make a significant contribution to the overall reproductive efficiency of the herd. It is therefore important to ensure that gilt management is at a level such that reproductive performance is maximized. In order to attain this aim the management should attempt to reduce the non-productive period from birth to first mating to a minimum while maintaining litter size at an acceptable level (9–10 piglets). Information presented in previous chapters suggests that a number of factors contribute to the realization of this objective:

(1) Inducing puberty attainment at approximately 160–170 days through the use of controlled boar introduction.
(2) Flushing the gilt for the one oestrous cycle between puberty and second heat.
(3) Mating at second oestrus.

These three major points should be built into a management system for gilts and probably this necessitates specialized housing for the gilt to facilitate effective oestrus testing and boar contact at the correct time. In the future if the use of oestrogen as a pubertal induction agent is developed this may considerably simplify the procedure particularly for the larger herd.

When the above system is used the replacement gilt should produce her first litter of 9–10 piglets at about 300–310 days. Since the normal age of gilts at first farrowing is in the region of 330–370 days, this represents a saving of around 75–100 kg of feed per gilt. Furthermore in terms of a 200 sow unit replacing 30 per cent of sows each year this represents a saving of up to 9 tonnes of feed per year.

15.2 The sow

Sow productivity has been defined as the number of piglets weaned/sow/year. This in turn is made up of two factors – litter size at weaning and the number of litters produced/sow/year.

LITTER SIZE

The average litter size at weaning which has been recorded in the UK is 8.6 piglets. However, in Chapter 8 it was suggested that a litter size at weaning of 9.6–10.5 piglets should be aimed for, by:

(1) selecting a hybrid strain of sow with a proven genetic ability to be prolific,
(2) providing optimum housing conditions throughout the sow's reproductive cycle,
(3) flushing between weaning and remating,
(4) maintaining as high a herd health and hygiene status as possible,
(5) low plane feeding throughout the *whole* of pregnancy, and
(6) detailed management control at farrowing and early lactation including the use of prostaglandin and a well designed farrowing crate.

In this way the number of piglets weaned per litter may be increased by as much as 25 per cent.

FARROWING INTERVAL

The farrowing interval represents the time for one complete reproductive cycle to occur, and therefore determines the number of litters produced by a sow per annum. It is composed of a 115 day pregnancy, a period of lactation, the duration of which is variable, and an interval from weaning to effective service. In order to reduce the farrowing interval, and hence increase the number of litters/sow/year, we may reduce either lactation length or the length of the period from weaning to effective service (i.e. 'empty days').

Reducing lactation length or early weaning is probably the most effective method of increasing the productivity of the sow. This is clearly shown in *Table 15.1.* However it should be reiterated that with our present state of knowledge very early weaning (below 3 weeks) will tend to be detrimental to sow productivity as a result of increased embryonic death. In view of this it is recommended that sows be weaned after lactations of 3–4 weeks.

Empty days – the period from weaning to effective service – have recently received much publicity since they represent the one time in the sow's reproductive cycle when she is non-productive (i.e. neither pregnant nor lactating). The magnitude of the problem becomes

Table 15.1. The effect of lactation length on the potential annual productivity of the sow

	Weaning age (weeks)			
	1	*3*	*6*	*8*
Gestation length (days)	115	115	115	115
Lactation length (days)	7	21	42	56
Weaning to remating interval (days)	5	5	5	5
∴ Farrowing interval (days)	127	141	162	176
∴ Litters/sow/year	2.87	2.59	2.25	2.07
∴ Piglets weaned/sow/year*	28.7	25.9	22.5	20.7

* Assuming a litter size at weaning of ten piglets.

Table 15.2. The effects of lactation length on 'empty days' in the sow

	Weaning age (weeks)		
	3–4	*5–6*	*7–8*
Average litters/sow/year*	2.2	1.9	1.7
∴ Average length of reproductive cycle (days)	166	192	215
Average length of pregnancy + lactation (days)	140	154	168
∴ Average 'empty days' per reproductive cycle†	26	38	47

* From M.L.C. (1975).
† Calculated as average length of reproductive cycle – (115 day pregnancy + mean lactation length).

evident when survey results on sow productivity are considered in terms of empty days. This has been done in *Table 15.2.* Clearly, 'empty days' severely reduce annual sow productivity. In *Table 15.2* the average number of empty days/cycle for all lactation lengths is 37, this being at least 30 days longer than the expected average interval from weaning to remating. As a result, annual productivity is depressed by at least 3–4 piglets per sow/year.

What then are the causes of such long periods from weaning to effective service? At their simplest they may be classified as follows:

(1) Sows failing to return to heat (anoestrus).
(2) Delay in return to heat.
(3) Failure to mate.
(4) Failure of conception or implantation.
(5) Abortion.
(6) Sows found non-pregnant at term.
(7) Delay in culling sows or introducing replacement gilts.

All these problems have been discussed in previous chapters and it is clear that most of them are unnecessary. It is suggested that a herd's average 'empty days' should be between 10 and 15 days or less and the following points are important in achieving this:

(1) A suitable individual sow monitoring scheme to follow the progress of individuals from weaning onwards. In the future computerized recording schemes should facilitate this.
(2) Culling decisions made accurately and early.
(3) Heat detection programmes followed through and executed rigorously.
(4) Service twice or possibly three times during each heat period using a boar of maximum fertility.
(5) Thorough checks for returns to service.
(6) Use of an accurate pregnancy diagnosis technique at strategic times throughout pregnancy.

This last point is at present somewhat limited by the techniques available and most equipment is not accurate enough early enough in pregnancy to be of much more use than checking for non returns. It is however encouraging to see a recent report by Macneil (1979) which describes a system being adopted in West Germany; this uses blood sampling and radioimmunoassay for progesterone for pregnancy diagnosis with very accurate results. If this becomes more widely used then it may prove a valuable tool for the pig farmer and for the veterinary surgeon.

15.3 The boar

Boar management has been discussed in detail in Chapters 11, 12 and 13. From this it is clear that the boar profoundly influences the apparent performance of the sow and any suboptimal fertility on the male side will have a significant adverse effect on the whole breeding herd. With this in mind the following points should be adhered to:

(1) Correct introduction procedure for young boars to ensure maximum libido and long working life.
(2) Correct working environment in terms of temperature, lighting and space.
(3) Regular veterinary inspection to monitor potential fertility problems.
(4) A detailed recording system to detect low fertility early.
(5) A thorough culling policy to remove boars as and when necessary before senility sets in.

The first point in this list is somewhat of a problem at present and is largely out of the hands of the commercial pig farmer. Most pig farmers, of course, buy in young boars at around 100 kg liveweight from either breeding companies or nucleus breeders by which time the damage has been done. It would be a step forward if in the future, pig farmers received more substantial assurance from the breeding companies that care in young boar rearing and management had been directed towards the enhancement of subsequent libido and fertility. At present no documentation exists to indicate the percentage wastage rate of young boars but undoubtedly this is a significant loss.

The second point, relating to the environment of the working boar, is also an area where more research effort should yield improvement for the commercial farmer. At present recommendations tend to be based on subjective judgement without sufficient evidence to define precise limits. It is hoped this situation is changing rapidly. In due course it may be that the importance of the 'on-farm working boar' will decrease as A.I. becomes a more practical and economic proposition. It has been shown in Chapter 14 that freezing techniques are now in commercial usage and if developments in this field continue this may add yet another dimension to the commercial pig breeding enterprise.

The conclusions drawn above and from preceding chapters show that there is much room for improvement in the productivity of the breeding herd. Advances in our understanding of reproductive processes, genetics, nutrition, environmental requirements and health have provided us with the tools with which we may greatly increase reproductive efficiency in the pig. The aim should be to disseminate this knowledge to the pig industry at large.

15.3 Reference

MACNEIL, F. (1979). *Pig Fmg* **27(9)**, 40–42

Tables of conversion factors

Length

1 foot	=	12 inches	=	0.3048 m
1 inch	=	2.54 cm	=	25.4 mm
1 yard	=	3 feet	=	0.9144 m
1 mile	=	1760 yards	=	1.6093 km

Area

1 sq foot	=	144 sq inches	=	0.0929 m^2
1 sq yard	=	9 sq feet	=	0.8361 m^2
1 acre	=	4840 sq yards	=	4047 m^2
1 sq mile	=	640 acres	=	259.0 hectares

Volume

1 cu inch	=	16.39 cm^3		
1 cu foot	=	1728 cu inches	=	0.0283 m^3
1 gallon (UK)	=	8 pints	=	4.546 litres
1 cu yard	=	27 cu feet	=	0.764 m^3
1 gallon (UK)	=	277.4 cu inches	=	1.201 gallons (US)
1 gallon (US)	=	37.8 litres	=	0.832 gallons (UK)

Mass

1 ounce	=	437.5 grains	=	28.35 g
1 pound	=	16 ounces	=	0.4536 kg
1 hundredweight	=	112 pounds	=	50.80 kg
1 ton (gross or long)	=	2240 pounds	=	1.016 tonnes
1 tonne	=	2205 pounds	=	0.984 gross tons
1 ton (short)	=	2000 pounds	=	0.907 tonnes
1 tonne	=	1000 kg		

Temperature

°*Fahrenheit* (°F)	°*Celsius* (°C)
10	−12.2
20	− 6.7
30	− 1.1
40	4.4
50	10.0
60	15.6
70	21.1
80	26.7
90	32.2
100	37.8
110	43.3
120	48.9
150	65.6
170	76.7
190	87.8
200	93.3
220	104.4

Energy

1 calorie (cal)	=	0.2388 joules (J)
1 joule (J)	=	4.1868 calories (cal)
1000 cal	=	1 kilocalorie (kcal)
1 kcal	=	4186.8 J
1 megajoule (MJ)	=	10^6 J
1 MJ	=	238.85 kcal

Glossary of terms

Abortion premature delivery of fetuses.

Anoestrous pig sow or gilt failing to show a heat for an indefinite period.

Anterior pituitary endocrine gland at the base of the brain which secretes and stores a multitude of different hormones many of which are involved in reproduction.

Artificial insemination a technique whereby fertilization is achieved without coitus taking place.

Blastocyst the late egg stage in mammalian development when cavitation has occurred.

Boar entire mature male pig.

Castration a technique effecting the inactivation of the gonads.

Catheter (A.I.) the artificial boar's penis used in artificial insemination.

Coitus sexual intercourse.

Colostrum the first milk of lactation containing a high protein content and a high content of immune globulins.

Conception the establishment of a successful pregnancy.

Conception rate the percentage of sows or gilts achieving successful pregnancy of those actually served.

Corpus luteum (pl. corpora lutea) a structure in the ovary, formed from the follicle following ovulation, which secretes progesterone and promotes successful pregnancy.

Creep an area of access to piglets separate from the sow lying area.

Creep feed supplementary feed given to piglets in a creep area while they are still suckling the sow.

Cystic ovary ovary having a proportion of abnormal cysts.

Diluent a substance used to dilute a sample of semen.

Ejaculation process of emission of semen from the penis.

Embryo stage of the developing offspring up to 4 weeks post coitum.

Embryo mortality rate percentage of fertilized ova dying while in the embryonic stage.

Empty days the number of days per reproductive cycle that a sow spends neither pregnant nor lactating.

Endogenous produced by an animal's own body resources (c.f. exogenous).

Ethology the study of animal behaviour.

Exogenous applied to an animal as an artificial additive to its own body resources (c.f. endogenous).

Extender a substance added to a sample of semen to extend its viable lifespan.

Farrowing parturition or delivery of piglets at the end of pregnancy.

Farrowing index the number of litters produced by each sow per year.

Farrowing interval the average time for a sow or a group of sows between one farrowing and the next.

Farrowing rate the number of sows actually farrowing as a percentage of sows served initially.

Fecundity the level of reproductive efficiency attained in terms of farrowing index.

Fertility the overall reproductive efficiency attained in terms of fecundity and prolificacy.

Fetus term used to describe offspring from the embryonic stage to parturition.

Flushing the technique of high plane energy feeding for a short time to augment ovulation rate.

Follicle a rapidly growing structure in the ovary which contains an ovum before ovulation.

Follicle stimulating hormone hormone secreted from the anterior pituitary gland which stimulates rapid growth of the follicles in the ovary.

FSH *see* Follicle stimulating hormone.

Gestation pregnancy.

Gestation length time from conception to parturition which is around 115 days in swine.

Gilt a female pig from birth to the time of conception of her second litter. Usually also applies to a prepubertal female.

Gilt replacement a gilt introduced into the breeding herd as a replacement for a cull sow.

Gonadotrophin an agent causing rapid growth and development of the follicles or ovaries, i.e. FSH and LH.

Heat period of sexual receptivity in the female.

Hormone a substance secreted by an endocrine gland in small amounts and carried in the blood stream to a target organ where it invokes a specific physiological response.

Hypothalamus a centre at the base of the brain which acts as a controller of a host of body functions – particularly those related to reproduction.

Hysterectomy the removal of the reproductive tract including the contents from the sow by a surgical technique.

Implantation the process of attachment of embryos to the uterine wall.

Infertility a state of reproductive malfunction that reduces overall efficiency of reproduction.

Infundibulum the terminal end of the oviduct which encloses the ovary.

Insemination introduction of semen into the female reproductive tract.

Intromission male penis penetrating the female vagina.

Lactation process of milk secretion.

Libido sexual drive.

Litter the total number of offspring born at one parturition.

Litter size the actual number of offspring (piglets) born or weaned per sow per farrowing.

Luteinizing hormone hormone responsible for the final maturation of ovarian follicles and the induction of ovulation.

Luteolytic an agent responsible for causing the regression of the corpora lutea.
Luteotrophic an agent responsible for causing the growth and maintenance of the corpora lutea.

Mating sexual intercourse.
Mating rate the percentage of sows or gilts exhibiting oestrus that are actually mated or served.
Milk ejection the process of milk letdown from the alveolar cells in the udder under the influence of the hormone oxytocin.
Mortality rate the percentage of individuals dying relative to the total individuals alive at the beginning of a given time period.

Natural service sexual intercourse between boar and sow.

Oestrus (heat) the exhibition of signs of sexual receptivity in the female under the influence of high levels of the hormone oestrogen in the blood stream.
Oestrous cycle the periodic appearance of oestrus (3 weeks) in unserved females.
Oestrus detection the regular and rigorous observance of sows or gilts for signs of oestrus.
Oestrus induction an agent or technique used to induce a sow or gilt to show oestrus at a time when she would not normally have done so.
Oestrous synchronization any technique which causes a group of females all to show oestrus at approximately the same time.
Oestrogen hormone secreted from the developing ovarian follicles.
Oogenesis process of growth and development of ova within the ovary.
Ovary female organ where ova are produced and shed, which is also responsible for the secretion of the hormones oestrogen and progesterone.
Overlying crushing of newborn piglets by the sow or gilt.
Oviduct (Fallopian tube) fine tubule which transports ova shed at ovulation to the uterus.
Ovulation process of shedding eggs from the ovary into the oviduct. Usually occurs in the middle of oestrus.
Ovulation rate the number of ova shed into the oviduct at ovulation.
Ovum (pl. ova) female egg cells containing half the genetic material necessary to initiate a new individual.
Oxytocin hormone secreted from the posterior pituitary gland which causes milk ejection and uterine contractions.

Parity the number of previous litters.
Parturition birth or farrowing.
Perinatal at or around parturition.
Perinatal loss a loss of piglets at or around the time of parturition.
Pineal a small endocrine gland, situated in the brain, which is thought to mediate light effects on reproduction.
Placenta the membranes providing the physiological link in the uterus between dam and offspring to facilitate the passage of nutrients and the removal of waste products.
Polytocous a species which produces more than one offspring at each parturition.
Posterior pituitary endocrine gland at the base of the brain releasing several hormones including oxytocin.
Precocious occurring earlier than anticipated.
Pregnancy the time between conception and parturition, i.e. gestation.
Prepubertal relating to the time before an animal has reached puberty.

Progesterone a steroid hormone secreted from the corpora lutea which can block the release of gonadotrophins from the anterior pituitary gland.
Prolactin a hormone from the anterior pituitary involved in the initiation and maintenance of lactation.
Prolificacy the level of reproductive output in terms of piglets produced per litter.
Prostaglandin a hormone secreted from the uterus which, among other functions, acts as a luteolytic agent, i.e. causes regression of the corpora lutea.
Puberty the time at which sexual and reproductive competence is reached.

Releasing hormone a hormone which initiates the release from another site in the body of a second hormone.
Reproductive cycle a complete sequence of events leading to the production of another generation of offspring, i.e. from service through pregnancy, lactation and weaning to the next service.

Semen fluid emitted from boar's penis at ejaculation containing sperm cells.
Seminiferous tubules site of spermatogenesis in the testes.
Service mating.
Service period the interval between weaning of the sow and remating.
Sperm male sex cells produced in the testes.
Spermatogenesis the process of sperm cell synthesis in the testes.
Sperm concentration the number of sperm cells present in a given quantity of semen.
Sperm motility the degree of movement observed by sperm cells in a given sample of semen.
Sow mature female pig. Normally in her second parity or reproductive cycle.
Sow productivity the level of output of piglets per sow per year.
Starvation death (of piglet) from exposure to cold and dampness.
Steroid hormones hormones of relatively simple molecular arrangement such as progesterone, oestrogen and testosterone.

Testes male sex organs where sperm cells are produced and stored.
Testosterone male hormone secreted by the testes, responsible for libido and secondary sexual characteristics.
Trophoblast layer of cells capable of rapid growth as in the outer layer of the developing blastocyst which implants into the uterine wall.

Uterus womb or female organ where the developing fetuses are nurtured.
Uterine involution the loss in weight and length of the uterus after farrowing.
Uterine endometrium the innermost layer of tissue within the uterine wall.

Weaner a weaned piglet in the early growth stage.
Weaning process of separation of suckling piglets from the sow.

Index